Fundamentals of Power System Transformers

"Electric machines" and "Transformers" are some of the most challenging electrical engineering courses offered to students. Their complexity arises from numerous prerequisites, a wide array of topics, and a combination of physics and mathematics, presenting students with significant challenges.

Fundamentals of Power System Transformers: Modeling, Analytics, and Operation acts as a stepping stone toward a deeper comprehension of the subject matter, resembling the content covered in a graduate-level course. The contents are condensed into two full chapters and four short chapters to provide a self-taught and self-sufficient book for students to solve all problems without the need for a computer.

Key features include:

- A variety of tests to prepare for entrance or employment exams
- Comprehensive coverage of transformers analysis, control, and protection
- Numerous problems and solutions with varying degrees of difficulty
- Problems that can be solved solely using a calculator, without dependence on any computer-based software
- Two-choice questions to reinforce readers' understanding of transformers concepts
- Explores not yet covered subjects including multi-winding auto-transformers, three-phase zigzag transformers, asymmetric and unbalanced three-phase transformers, special transformers, and transformer control

This book is aimed at graduate students taking classes in electrical engineering and serves as a valuable reference for researchers and industry professionals interested in emerging technologies and innovations in power system transformers.

Fundamentals of Power System Transformers

Modeling, Analysis, and Operation

Mostafa Eidiani and Kumars Rouzbehi

CRC Press
Taylor & Francis Group
Boca Raton London New York

CRC Press is an imprint of the
Taylor & Francis Group, an **informa** business

Designed cover image: Shutterstock

First edition published 2025
by CRC Press
2385 NW Executive Center Drive, Suite 320, Boca Raton FL 33431

and by CRC Press
4 Park Square, Milton Park, Abingdon, Oxon, OX14 4RN

CRC Press is an imprint of Taylor & Francis Group, LLC

ISBN: 9781032881751 (hbk)
ISBN: 9781032883977 (pbk)
ISBN: 9781003537564 (ebk)

DOI: 10.1201/9781003537564

Typeset in Times
by codeMantra

Access the Support Material: Routledge.com/9781032881751

Eidiani *dedicated this book to his beloveds, Elham and Hosna.*

Rouzbehi *presented this book to his late mother, who continues to be the guiding light in every page turned.*

Jalāl al-Dīn Muḥammad Rūmī *As you start to walk on the way, the way appears.*

Contents

Abstracts

CHAPTER 1

An electric **transformer** is a passive AC component that transfers electrical energy from one electrical circuit to another or multiple circuits. Its main function is to change voltage and current levels. To understand the fundamental principles of transformers, a conceptual model called the **ideal transformer** is used. In fact, an ideal transformer is a theoretical device in electrical engineering and physics, designed to simplify the analysis of real-world transformers. A transformer free from all sorts of losses is known as an ideal transformer. This means that this imaginary transformer has no core loss, ohmic resistance, leakage flux, or hysteresis, and eddy current losses. In addition, the ideal transformer core has infinite permeability, meaning less magnetizing current is required to magnetize the core. The leakage flux in an ideal transformer is zero, meaning the entire flux induced in the core links with both the primary and secondary windings. Therefore, an ideal transformer has 100% efficiency.

In the real world, transformers inevitably have limitations and losses compared to their ideal counterparts such as non-zero losses, copper losses, core losses, leakage flux, and reduced efficiency. While the ideal transformer model provides a conceptual foundation, considering the realities of non-idealities in real transformers is essential for practical application.

In the following, firstly, necessary formulas are recalled, and then ideal transformers and autotransformers with multiple windings are presented. Then the real single-phase transformer is analyzed. Concepts and solved problems are also presented for various transformer connections, including series–parallel configurations. At the end, in addition to the two-choice questions (Yes/No), a larger number of solved problems are given.

CHAPTER 2

Imagine a scenario where you have three single-phase transformers, each operating on its independent AC circuit. Now, visualize connecting these transformers in a precise configuration to harness them as a unified three-phase power source. This configuration essentially embodies a three-phase transformer.

Each three-phase transformer has primary windings connected to a three-phase source and secondary windings delivering the transformed power. Depending on the desired voltage and current levels, the windings can be connected in one of two main ways: star (Y) or delta (Δ). In a star connection, the ends of all three wires (phase wires) are connected together, while in a delta connection, they are left open. Three-phase transformers are widely used in various industries for diverse applications such as power transmission, sub-transmission, distribution, and industrial equipment. Compared to single-phase transformers, they require more complex design, manufacturing, and installation, making them more expensive. For optimal performance, regular maintenance is crucial.

To read this chapter, you must have read the previous chapter. The first section of this chapter discusses the fundamental concepts of three-phase systems. Next, different types of ideal three-phase transformers are presented. In addition to the famous star and delta connections, the zigzag connection is also investigated. Following the discussion on transformer connections, some important application-oriented subjects are explored as well. A significant number of solved problems are included at the end of the chapter, in addition to the two-choice questions (yes/no).

CHAPTER 3

The world of transformers extends far beyond the metal giants that convert voltage in power plants and substations. While these important elements in the power grid step up or down voltage levels, there are specialized transformers, each designed for a unique purpose. This chapter delves into the diverse world of special transformers revealing their critical role in various applications.

Instrument transformers, encompassing current transformers (CTs) and voltage transformers (VTs) [or potential transformers (PTs)], play a crucial role in electrical measurements. Phase-shifting transformers (PSTs) manipulate the phase angle of electrical signals, which is vital for power flow control within the grid. Giant High-Voltage Direct Current (HVDC) transformers, on the other hand, bridge the gap between AC and DC power transmission, enabling efficient long-distance power transfer.

Power distribution transformers are local power delivery transformers that step down high voltage from transmission lines to usable voltage levels for homes and businesses. Furnace transformers provide the high voltage required for electric furnace operation, while isolation transformers create electrically isolated circuits for safety or to eliminate ground loops.

Dry-type transformers, unlike their oil-filled counterparts, offer a cleaner and more environmentally friendly alternative. Welding transformers provide the high current and low voltage necessary for the intense heat required for welding processes.

Grounding transformers provide a crucial path for fault currents, enhancing safety and stability. Rotary transformers enable power and signal transfer between rotating parts for instance in a brushless excitation system of generators. Variable-frequency transformers (VFTs) are controllable bidirectional transmission devices that can transfer power between asynchronous networks.

In the following, you will read more details about planar transformers, their construction, advantages, and specific applications in power systems.

CHAPTER 4

This chapter is focused on the control of transformers, especially three-phase transformers. Voltage regulation and tap changers are the most important elements of voltage control in transformers. No-load tap changers (NLTC) and on-load tap changers (OLTC) are the two tap changer setup methods that are reviewed in this chapter. By examining thermal considerations in transformers, common and advanced methods of temperature control are briefly mentioned. It has also been shown that continuous load monitoring and remote control and monitoring allow operators to identify situations where the load is approaching rated capacity and take corrective action before overloading occurs. In the end, in addition to reviewing the cooling methods of transformers, control methods based on power electronics and FACTS (flexible AC transmission system) devices and their interaction with transformers have been reviewed.

CHAPTER 5

In this chapter, another fundamental practical concept needed by electrical engineers and students about industrial transformers is reviewed. First, routine tests, startup tests, and typing tests are reviewed and the differences and similarities between them are reviewed. Then, the importance of diagnostic test methods is emphasized for early identification of possible issues, avoiding expensive outages, and failures, and increasing the life of transformers. In the following, overcurrent protection and differential protection are discussed.

How to detect the presence of gas and abnormal current in transformers by using two sudden pressure relays (SPRs) and Buchholz relays and their advantages and disadvantages are discussed. Then critical safety considerations for transformers and the understanding of the factors affecting the transient life and reliability of transformers are reviewed. In the end, with the two-choice question-answers of this chapter, we will check the performance of the readers' understanding of this section.

CHAPTER 6

In this chapter, you will learn about the emerging technologies and frontiers of knowledge in transformers. The first topic is the concept of digital transformers, which are ordinary transformers combined with sensors, communication modules, and intelligent controllers. Then there is a list of exciting developments in transformers, including amorphous metal cores, advanced insulating fluids, Nano-crystalline materials, digital twins, solid-state transformers (SSTs), ester coolers, and more. In the following, the role of transformers in integration with smart grids has been examined. Then the advantages and disadvantages of high-temperature superconducting transformers (HTS) are presented. It is further shown that modular transformers offer a flexible approach to power delivery, are more adaptable, easier to maintain, and offer a wider range of applications. Finally, the use of artificial intelligence in the advanced analysis of transformer data is reviewed. In the end, with the two-choice question-answers of this chapter, we will check the performance of the readers' understanding of this section.

Foreword

The field of electrical power engineering is known for its complexity and critical importance to modern society. At the heart of this realm lies the power system transformer, a vital component whose role is indispensable in transmitting and distributing electrical energy. As the global energy demands surge and the quest for sustainable and resilient power systems intensifies, a profound understanding of transformer technology becomes more crucial.

Fundamentals of Power System Transformers: Modeling, Analysis, and Operation is a timely and definitive work that addresses this need with exceptional clarity and depth. This book, authored by two leading experts with over 20 years of experience in both academia and industry, offers a comprehensive exploration of transformer theory, enriched with practical insights into their operation and performance. It stands as a testament to the authors' profound expertise and commitment to advancing knowledge in electrical power engineering.

The text is meticulously structured to cover the full spectrum of topics essential to mastering transformer technology. Beginning with the fundamental principles and the core characteristics of transformers, it progresses through detailed modeling and analytical methods. Each chapter is crafted to build systematically upon the preceding material, ensuring a coherent and cumulative educational experience.

One of the notable strengths of this book is its integration of theoretical constructs with practical applications. The inclusion of real-world examples serves to bridge the gap between abstract concepts and their tangible implementations in the field. This approach not only enhances comprehension but also demonstrates the pivotal role of transformers in various electrical power systems.

Furthermore, this volume does not merely dwell on established knowledge but also addresses contemporary challenges and emerging trends. The discussions on the integration of renewable energy sources, the advancements in smart grid technologies, and the imperative of enhancing energy efficiency are pertinent. These sections provide readers with insights into the future trajectory of transformer technology and its evolving role in a modern, sustainable energy landscape.

For students of electrical engineering, this book will serve as an invaluable educational resource, guiding them through the complexities of transformer technology with clarity and precision. Practicing engineers will find it a thorough reference that offers both a refresher on foundational principles and a source of new methodologies to enhance their professional practice. Researchers, too, will appreciate the solid theoretical groundwork laid here, which can serve as a springboard for further innovation and exploration in the field.

Fundamentals of Power System Transformers: Modeling, Analysis, and Operation stands as a seminal work, seamlessly blending rigorous theoretical analysis with practical matters. This indispensable volume is a vital addition to the professional library of anyone involved in the study or practice of electrical power engineering. As you explore the profound insights contained within these pages, you will undoubtedly find this book to be an illuminating and enriching contribution to your professional development and a deeper understanding of power system transformers.

Professor Hassan Modir Shanechi
Electrical and Computer Engineering Department
Illinois Institute of Technology, Chicago, USA

Preface

In terms of comprehension, "Electric machines" and "Transformers" are some of the most challenging electrical engineering courses offered to junior students. Their complexity arise from numerous prerequisites, a wide array of topics, and a combination of physics and mathematics, presenting students with substantial challenges.

Each chapter within this book acts as a stepping stone toward a deeper comprehension of the subject matter, resembling the content covered in an undergraduate-level course. We have condensed all the necessary contents into two comprehensive chapters and four short chapters. The majority of universities worldwide cover this material in a semester.

In Chapter 1, first, the necessary formulas are recalled, and then, ideal transformers and autotransformers with multiple windings are presented. Then, the real single-phase transformer is analyzed. Concepts and solved problems are also presented for various transformer connections, including series–parallel configurations.

Chapter 2 first discusses the basic concepts of three-phase power systems. Next, different types of ideal three-phase transformers are examined. In addition to the famous star and delta connections, as well as others, the zigzag connection is also fully investigated. In the following, ideal and real three-phase transformers have been analyzed.

Chapter 3 discusses the special transformers including current and voltage transformers, phase shifting, HVDC, distribution, furnace, isolation, and dry-type transformers.

Chapter 4 discusses the fundamental concepts of transformer control. Transformer control classically refers to the methods and systems employed to manage and regulate the operation of a transformer in a power system. The control of a transformer is essential to ensure it operates within specified limits, meets the required performance criteria, and responds appropriately to variations in the electrical load or other system conditions.

Chapter 5, covering transformer testing, diagnosis, and protection, and Chapter 6, focusing on emerging technologies in transformers, have been condensed to provide a concise introduction to these topics and help you become acquainted with them.

University students can use this book to deepen their knowledge, improve their study speed, and review the most critical lessons of the subject to prepare for the course exam. In addition, this book is a valuable resource for both undergraduate (BEE) and graduate (MEE) students and professors in the field of electrical engineering. Additionally, it serves as a convenient reference for electrical engineers working in industries providing quick access to essential formulas and concepts.

This book contains an explanation sheet that can be used to reread and summarize needed materials to solve the related problems. Figure slides are also available on the Routledge website (www.routledge.com/9781032036533).

Other advantages associated with delving into this book are as follows:

- This book contains a variety of tests to prepare for entrance or employment exams.
- Electrical engineers dealing with transformer analysis, control, and protection can find almost everything they need in this book.
- This book contains a significant number of both difficult and easy problems and solutions.
- Readers can solve problems presented in this book solely using a calculator, without dependence on any computer-based software.
- This book provides transformers concepts by studying two-choice questions.

- This book coverers not yet covered subjects (by other similar books), like multi-winding auto-transformers, three-phase zigzag transformers, asymmetric and unbalanced three-phase transformers, special transformers, transformer control, and …

In the end, we had a great time writing this book, and we truly hope you enjoy reading it as much as we enjoyed creating it!

About the Authors

Mostafa Eidiani (StM'98, SM'16) earned a B.S. (with distinction) and M.Eng. degree in Electrical Engineering from Ferdowsi University of Mashhad, Iran, in 1995 and 1997, respectively, as well as his Ph.D. degree from the Science and Research Branch of Islamic Azad University, Tehran, Iran, in 2004. In 2016, he was elevated as an IEEE senior member, and in 2017, he was elected as the director of the Iranian Association of Electrical and Electronics Engineers (Khorasan Branch). He was promoted from assistant professor to associate professor in 2016. His research interests include renewable energy integration, power system control, transient and voltage stability, power system simulation, and DIgSILENT PowerFactory simulations and analysis.

He has authored or co-authored 11 technical books, 9 chapter books, 40 journal papers, and 110 technical conference proceedings. In addition, he has conducted more than 30 research projects with Iranian power companies. He is an associate editor for the *IET Journal of Engineering (JOE)* and *Journal of Electrical Engineering & Technology (JEET)*, and an editorial board member for the *International Journal of Applied Power Engineering (IJAPE)* and nine other journals. He has been a board member of Khorasan Electric Generation Company.

He is the author of *Fundamentals of Power Systems Analysis I: Problems and Solutions* and *Advanced Topics in Power Systems Analysis: Problems, Methods, and Solutions* published by Taylor & Francis Group.

He was included in the list of the World's Top 2% Scientists List 2024, published by Stanford University.

Kumars Rouzbehi (Senior Member, IEEE) received his Ph.D. degree in Electric Energy Systems from the Technical University of Catalonia (UPC), Barcelona, Spain, in 2016. Prior to this, he was an academic staff member at the Islamic Azad University (IAU), Iran, from 2002 to 2011. In parallel with teaching and research at the IAU, he was the CEO of Khorasan Electric and Electronics Research Company, from 2004 to 2010. From 2017 to 2018, he was an associate professor at the Loyola Andalucía University, Seville, Spain. In 2019, he joined the Department of System Engineering and Automatic Control at the University of Seville, Spain. He is the patent holder for AC grid synchronization of voltage source power converters and has contributed to over 100 technical publications, including books, book chapters, journal papers, and technical conference proceedings. Professor Rouzbehi has been a TPC Member of the International Conference on Electronics, Control, and Power Engineering (IEEE.ECCP), since 2014; a scientific board member of the (IEA) International Conference on Engineering and Management, since 2015; and a TPC Member of COMPEL 2020.

He is an associate editor of the *IEEE Systems Journal, IET Generation, Transmission and Distribution, IET Renewable Power Generation, High Voltage (IET)*, and *IET Systems Integration*. He received the Second Best Paper Award 2015 from the *IEEE Power Electronics Society, IEEE Journal of Emerging and Selected Topics in Power Electronics*.

He is a co-author of *Active Filter Design (in Persian)* and *Advanced Topics in Power Systems Analysis: Problems, Methods, and Solutions* published by CRC/Taylor & Francis Group.

1 Single-Phase Transformers

Part One: Lesson Summary

1.1 RECALLING FUNDAMENTAL RELATIONSHIPS

In this section, single-phase power-related equations are revisited [1].
Phasor representation would be:

$$\hat{V} = V\angle\theta_v, \quad \hat{I} = I\angle\theta_i \tag{1.1}$$

The angle between voltage and current:

$$\varphi = \theta_v - \theta_i \tag{1.2}$$

The real or active power (Watt or W):

$$P = VI\cos(\varphi) \tag{1.3}$$

The imaginary or reactive power (Volt-Ampere reactive or VAr):

$$Q = VI\sin(\varphi) \tag{1.4}$$

The complex power (Volt-Ampere or VA):

$$S = P + jQ = \hat{V}\hat{I}^* = VI\angle\varphi = |S|\angle\varphi = (\sqrt{P^2+Q^2})\angle\varphi \tag{1.5}$$

The apparent power or the magnitude of the complex power (VA):

$$|S| = \sqrt{P^2+Q^2} \tag{1.6}$$

In the case of an impedance load, the following equations and the two similar triangles in Figure 1.1 can be considered [2–5].

$$P = VI\cos(\varphi) = RI^2, \quad Q = VI\sin(\varphi) = XI^2 \quad |S| = VI = |Z|I^2 = \frac{V^2}{|Z|} \tag{1.7}$$

The power triangle is obtained by multiplying the impedance triangle by I^2.

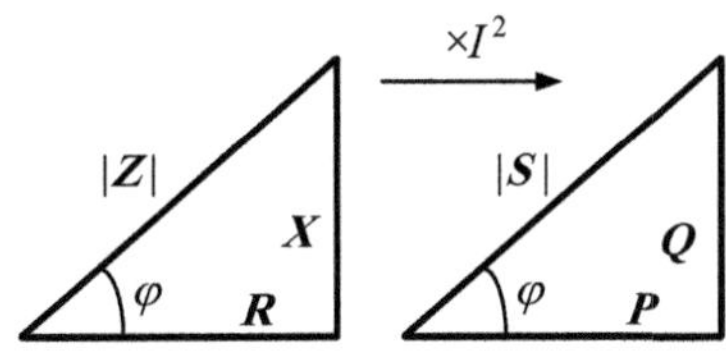

FIGURE 1.1 Impedance and power triangles.

DOI: 10.1201/9781003537564-1

Note: Imaginary (or reactive) power is a double-frequency sinusoid with **zero average value** and an **amplitude *Q*** given by equation (1.4). In fact, reactive power is the part of complex power (equation 1.5) that corresponds to the storage and retrieval of energy, rather than its consumption.

1.2 IDEAL TRANSFORMER

Equations (1.8) and (1.9) define the voltage and current relationships in an ideal transformer. Two single-phase transformers are depicted in Figures 1.2 and 1.3. Figure 1.3 shows that the coils in parallel are in phase, regardless of whether they are oriented vertically or horizontally.

Equations (1.8) and (1.9) hold true for all multi-winding transformers (Figure 1.4) and three-phase transformers, where the parallel windings are in phase. Equation (1.8) demonstrates the direct correlation between voltage and the coil's number of turns. The direction of winding is indicated by the **dot marks** in Figures 1.2–1.4.

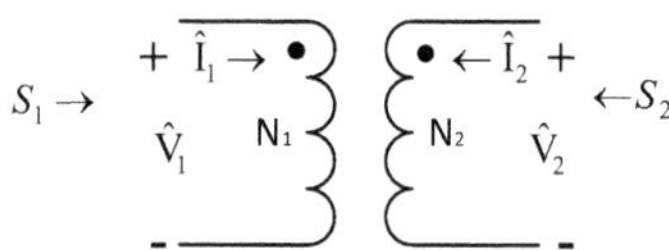

FIGURE 1.2 An ideal two-winding transformer in the general form.

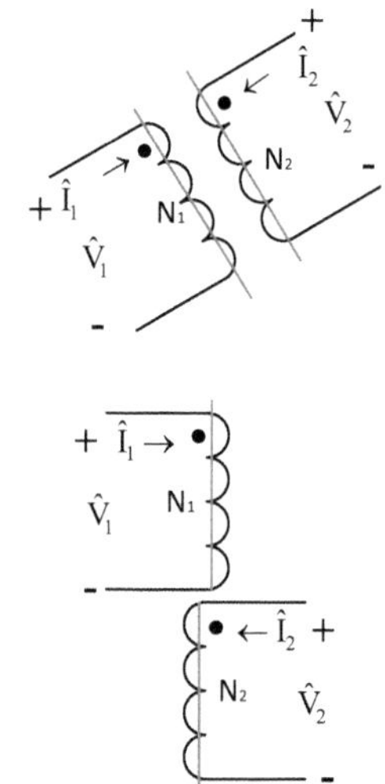

FIGURE 1.3 Two parallel coils in-phase (same as Figure 1.2).

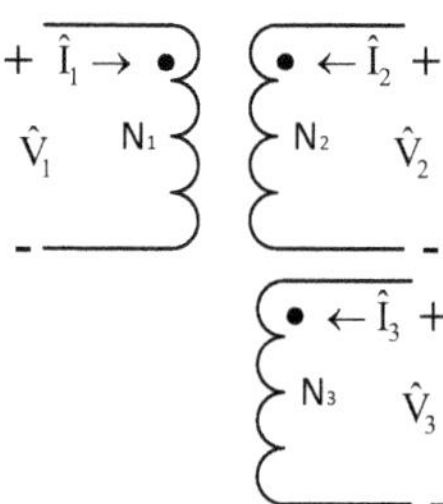

FIGURE 1.4 An ideal three-winding transformer in the general form.

According to equation (1.9), for in-phase windings in an ideal transformer, the magnetic motive force (mmf) is zero. Thus, equation (1.9) indicates that the total number of turns in each coil, multiplied by the input current at the dot marks on the coil, equals zero. This equation holds true for all coils that are in phase.

$$\frac{\hat{V}_1}{\hat{V}_2} = \frac{N_1}{N_2} \Rightarrow V \propto N \tag{1.8}$$

$$N_1\hat{I}_1 + N_2\hat{I}_2 = 0, \quad \text{or:} \quad \sum_{i=1}^{n}\left(N_i\hat{I}_i\right) = 0 \tag{1.9}$$

Figure 1.5 and equation (1.10) are usually used for two-winding transformers. The difference between Figures 1.2 and 1.5 lies in the direction of the load current.

$$\frac{\hat{V}_1}{\hat{V}_2} = \frac{N_1}{N_2} = a = \frac{\hat{I}_2}{\hat{I}_1} \tag{1.10}$$

Ideal transformers meet the following requirements:

- Their copper and core losses are zero.
- There is no saturation effect.
- Their total input and total output powers are equal. For instance, in a two-winding transformer, this means (See Figure 1.2):

$$S_1 = \hat{V}_1\hat{I}_1^* = \left(\frac{N_1}{N_2}\hat{V}_2\right)\left(-\frac{N_2}{N_1}\hat{I}_2^*\right) = -\hat{V}_2\hat{I}_2^* = -S_2 \Rightarrow S_1 + S_2 = 0 \tag{1.11}$$

Or (See Figure 1.5):

$$S_{\text{in}} = \hat{V}_1\hat{I}_1^* = \hat{V}_2\hat{I}_2^* = S_{\text{out}}, \quad P_{\text{in}} = P_{\text{out}}, \quad Q_{\text{in}} = Q_{\text{out}} \tag{1.12}$$

Equation (1.12) is also true for multi-winding ideal transformers. It is easy to calculate the impedance seen from the primary side or the impedance transferred to the transformer primary. From Figure 1.5 and equation (1.10), we have:

$$Z_{\text{in}} = \frac{\hat{V}_1}{\hat{I}_1} = \frac{a\hat{V}_2}{\left(\frac{\hat{I}_2}{a}\right)} = a^2\frac{\hat{V}_2}{\hat{I}_2} = a^2 Z_{\text{Load}} = \left(\frac{N_1}{N_2}\right)^2 Z_{\text{Load}} \tag{1.13}$$

In other words, the impedance passes through the transformer with a conversion ratio to the power of 2.

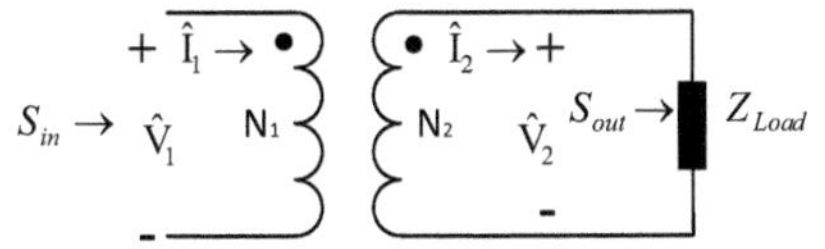

FIGURE 1.5 An ideal two-winding transformer in the usual form.

1.3 IDEAL AUTOTRANSFORMER

A transformer with an electrical connection between its primary and secondary windings is called an autotransformer, or simply an "auto-trans". The term "autotransformer" means a single-winding transformer type. In fact, the prefix "auto" denotes a single coil that runs independently rather than automatically. Figure 1.6 shows a two-winding autotransformer made from a two-winding transformer, while Figure 1.7 shows a three-winding autotransformer made from a three-winding transformer. In Figures 1.6 and 1.7, the equations (1.8) and (1.9) are valid.

The ideal transformer equation is unaffected by the electrical connection. The input and output of the autotransformer differ from those of the primary transformer, as shown in Figure 1.6. We have:

$$\hat{V}_A = \hat{V}_1 + \hat{V}_2, \quad \hat{I}_A = \hat{I}_1, \quad \hat{V}_B = \hat{V}_2, \quad \hat{I}_B = \hat{I}_2 - \hat{I}_1 \tag{1.14}$$

From equations (1.8), (1.9), and (1.14) we have:

$$\frac{\hat{V}_A}{\hat{V}_B} = \frac{\hat{V}_1 + \hat{V}_2}{\hat{V}_2} = \frac{N_1}{N_2} + 1, \quad \frac{\hat{I}_B}{\hat{I}_A} = \frac{\hat{I}_2 - \hat{I}_1}{\hat{I}_1} = -\left(\frac{N_1}{N_2} + 1\right) \tag{1.15}$$

Now, the apparent power ratio of the auto-trans is compared to that of the primary transformer. If the auto-trans is step-down, as shown in Figure 1.6, the relation (1.16) applies, and if it is step-up, the relation (1.17) is established.

$$\left|\frac{S_A}{S_1}\right| = \frac{V_A I_A}{V_1 I_1} = \frac{V_A}{V_1} = \frac{V_1 + V_2}{V_1} = 1 + \frac{N_2}{N_1} \tag{1.16}$$

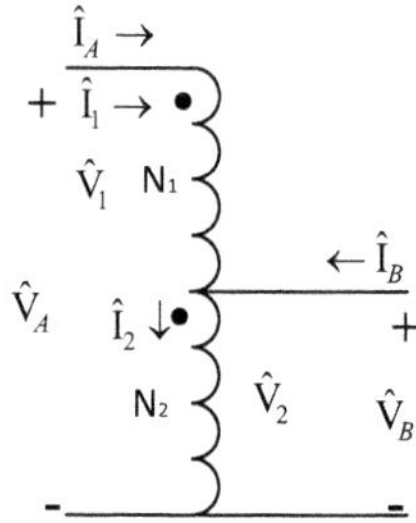

FIGURE 1.6 Ideal two-winding autotransformer.

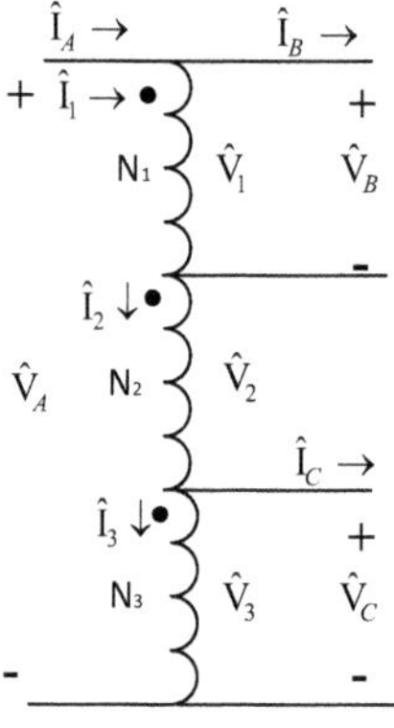

FIGURE 1.7 Ideal three-winding autotransformer.

$$\left|\frac{S_A}{S_1}\right| = \frac{V_A I_A}{V_1 I_1} = \frac{I_A}{I_1} = \frac{I_1 + I_2}{I_1} = 1 + \frac{N_1}{N_2} \tag{1.17}$$

It is simple to demonstrate that an auto-trans is a transformer with a change in conversion ratio using equation (1.15). Insulation is a critical safety concern because of the electrical connection between the auto-trans' primary and secondary windings. (See Figure 1.7) We have:

$$\frac{\hat{V}_1}{\hat{V}_2} = \frac{N_1}{N_2}, \quad \frac{\hat{V}_1}{\hat{V}_3} = \frac{N_1}{N_3} \Rightarrow V \propto N \tag{1.18}$$

$$N_1 \hat{I}_1 + N_2 \hat{I}_2 + N_3 \hat{I}_3 = 0 \tag{1.19}$$

KVL and KCL relations are, of course, established for every circuit.

1.4 EQUIVALENT CIRCUIT MODEL OF A REAL TRANSFORMER

1.4.1 Accurate Model

In a real transformer, the electric and magnetic characteristics must be reflected in the equivalent circuit. Characteristics that depend on current are modeled in series, while those that depend on voltage are modeled in parallel. Here, copper or ohmic losses (in the form of series resistance), leakage flux (in the form of series inductance), core losses (in the form of parallel resistance), and magnetic (or mutual) flux (in the form of parallel inductance) should be added to the ideal transformer model. An accurate model of a single-phase real transformer (or a single-phase form of a symmetrical three-phase transformer) is shown in Figure 1.8.

Where:

R_1: Primary winding resistance, the copper loss component of the primary winding
R_2: Secondary winding resistance, the copper loss component of the secondary winding
X_1 or X_{l1}: Primary winding reactance, leakage flux component of the primary winding
X_2 or X_{l2}: Secondary winding reactance, leakage flux component of the secondary winding
$\hat{V}_1, \hat{I}_1$: Primary side source voltage and current
$\hat{V}_2, \hat{I}_2$: Secondary side load voltage and current
$\hat{I}_\varphi \left(= \hat{I}_c + \hat{I}_m\right)$: Excitation current or no-load current in the transformer
$\hat{I}_c$: Core-loss (hysteresis and eddy) component of the excitation current
$\hat{I}_m$: Magnetizing component of the excitation current
R_c: Equivalent resistance of the core-loss component
X_m: Equivalent reactance of the magnetizing component

Working with the equivalent circuit in Figure 1.8 is a bit difficult due to the presence of an ideal transformer in the middle of the circuit. It is better to transfer the secondary part to the primary part with the help of equation (1.13). Therefore, we reach the equivalent circuit in Figure 1.9.

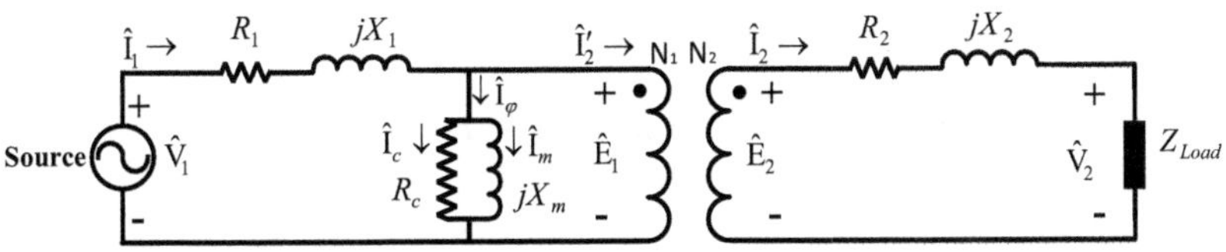

FIGURE 1.8 The accurate model of a single-phase real transformer.

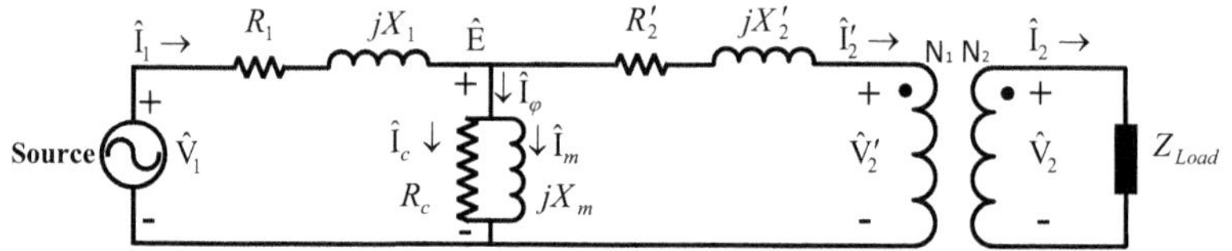

FIGURE 1.9 The accurate model of a single-phase real transformer which is referred to as the primary.

Where:

$$R_2' = \left(\frac{N_1}{N_2}\right)^2 R_2 \quad \text{Secondary winding resistance refers to the primary} \tag{1.20}$$

$$X_2' = \left(\frac{N_1}{N_2}\right)^2 X_2 \quad \text{Secondary winding reactance refers to the primary} \tag{1.21}$$

$$\hat{V}_2', \hat{I}_2' \quad \text{Secondary side load voltage and current refer to the primary,} \quad \frac{\hat{V}_2'}{\hat{V}_2} = \frac{N_1}{N_2} = \frac{\hat{I}_2}{\hat{I}_2'} \tag{1.22}$$

Now, if we assume that the load information is known, the remaining values of voltage and current can be calculated.

With a decent approximation, the parallel branch can be removed from the middle of the circuit. In this case, it comes to the approximate circuit of the transformer.

1.4.2 Approximate Model

Equivalent transformer impedance (total transformer impedance) is defined as follows.

$$Z_{eq} = R_{eq} + jX_{eq} = (R_1 + R_2') + j(X_1 + X_2') \tag{1.23}$$

Four types of approximate equivalent circuits can be seen in Figure 1.10 (a, b, c, d).

The cases of using approximate models are related to the working conditions of the transformer.

a. (a): An approximate model (a) is used when the load voltage is known.
b. (b): An approximate model (b) is used when the source voltage is known.
c. (c): An approximate model (c) is used when core parameters are not important.
d. (d): An approximate model (d) is usually used in simplified power systems.

1.4.3 Power Equations

In the accurate model of the transformer, the active power equations are written as follows.

$$\text{Input power or power source}: P_{in} = V_1 I_1 \cos\varphi_1 \tag{1.24}$$

$$\text{Output power or load power}: P_{out} = V_2 I_2 \cos\varphi_2 = V_2' I_2' \cos\varphi_2 \tag{1.25}$$

$$\text{Primary copper or ohmic losses}: P_{cu1} = R_1 I_1^2 \tag{1.26}$$

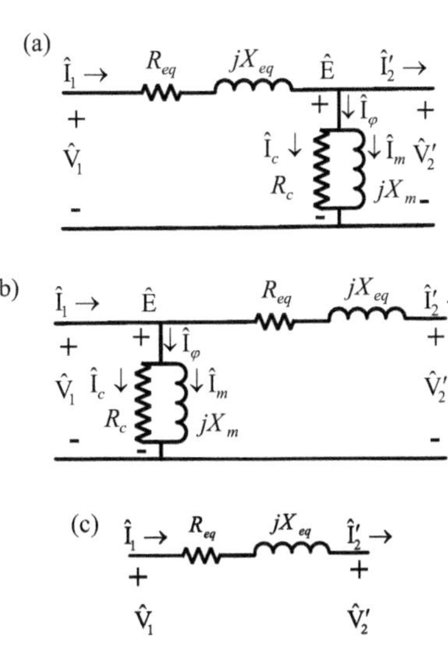

FIGURE 1.10 The approximate models of a single-phase real transformer.

$$\text{Secondary copper or ohmic losses}: P_{\text{cu}2} = R_2 I_2^2 = R_2' I_2'^2 \tag{1.27}$$

$$\text{Core losses}\left(\text{hysteresis and eddy}\right): P_c = R_c I_c^2 = \frac{E^2}{R_c} \tag{1.28}$$

$$\text{Total losses}: P_{\text{loss}} = P_{\text{cu}1} + P_{\text{cu}2} + P_c = P_{\text{in}} - P_{\text{out}} \tag{1.29}$$

$$\text{Efficiency}: \eta = \frac{P_{\text{out}}}{P_{\text{in}}}, \quad \%\eta = 100\frac{P_{\text{out}}}{P_{\text{in}}} \tag{1.30}$$

$$\text{Note}: R_1 \simeq R_2', \quad X_1 \simeq X_2', \quad R_c >> R_1, \quad X_m >> X_1 \tag{1.31}$$

1.5 NOMINAL VALUES OF THE TRANSFORMER

Nominal values (or nameplate values) are the values for which the transformer is designed. These values are written on the nameplate of the transformer. Usually, each device is designed and manufactured to work with its nominal values.

$V_n\left(V_{n1}, V_{n2}\right)$: Nominal transformer voltage (Primary and Secondary)

$I_n\left(I_{n1}, I_{n2}\right)$: Nominal transformer current (Primary and Secondary)

$S_n = (V_{n1})(I_{n1}) = (V_{n2})(I_{n2})$: Nominal transformer power

1.6 VOLTAGE REGULATION

Voltage regulation is the change in voltage at a point in response to a change in load. Voltage regulation should not be confused with voltage drop, as voltage drop calculates voltage at two points during a load change.

$$\%\text{Voltage Regulation} = \%V_R = (100)\frac{V_{2nl} - V_{2fl}}{V_{2fl}} = (100)\frac{V'_{2nl} - V'_{2fl}}{V'_{2fl}} \tag{1.32}$$

Where:

V_{2nl} (No-Load): Load point voltage when the load is disconnected

V_{2fl} (Full-Load): Load point voltage when the load is connected

In the approximate model of Figure 1.10c, the voltage regulation can be written as follows:

$$\%\ V_R = (100)\frac{V_1 - V_2'}{V_2'} \tag{1.33}$$

With a decent approximation, if the power of the load is known, the relation (1.33) can be written as follows:

$$\%V_R = (100)\frac{RP \pm XQ}{(V_2')^2}\ (+: \text{lagging load})(-: \text{leading load}) \tag{1.34}$$

1.7 PERUNIT SYSTEM

The perunit is conceptualized to eliminate the effect ideal transformer. Parameters are perunited by dividing them into parameter bases. Parameters and base parameters:

$$\underbrace{S^{VA}, P^{W}, Q^{VAr}}_{S_b^{VA}}, \underbrace{Z^{\Omega}, R^{\Omega}, X^{\Omega}}_{Z_b^{\Omega}}, \underbrace{Y, G, B}_{Y_b^{\mho}}, \underbrace{V^{V}}_{V_b^{V}}, \underbrace{I^{A}}_{I_b^{A}} \tag{1.35}$$

1.7.1 Perunit System for Single-Phase

V_b, phase voltage and S_b, single-phase power are the main bases and are usually known. All other bases can be calculated using the following equations. The nominal values are usually equal to these base values. It is assumed that the system's maximum rated power is S_b. In the first generator, the voltage was V_b, which then passes through the transformers with a conversion ratio.

$$Z_b = \frac{V_b^2}{S_b},\quad I_b = \frac{V_b}{Z_b},\quad Y_b = \frac{1}{Z_b} \tag{1.36}$$

1.7.2 Perunit System for Three-Phase

V_b (line-to-line voltage) and S_b (three-phase power) are the main bases and are usually known. All other bases can be calculated using the following equations:

$$Z_b = \frac{V_b^2}{S_b},\quad I_b = \frac{V_b}{\sqrt{3}Z_b},\quad Y_b = \frac{1}{Z_b} \tag{1.37}$$

Three-phase perunit systems eliminate coefficients 3 and $\sqrt{3}$ in power and voltage, such as $P^{pu} = V^{pu}I^{pu}\cos\varphi$ for three-phase power.

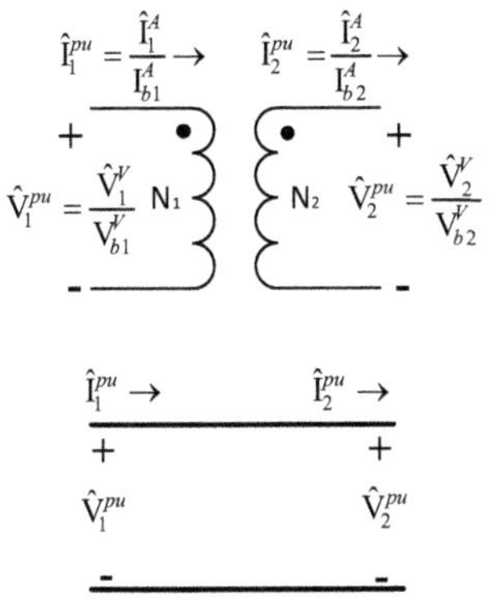

FIGURE 1.11 Remove the ideal transformer in the perunit mode.

1.7.3 Eliminating the Perunit Ideal Transformer

Figure 1.11 shows how to remove the ideal transformer in the perunit mode.

The point of perunit transformer is that the bases on both sides of the transformer are transferred with the transformation ratio.

$$\frac{V_{b1}}{V_{b2}} = \frac{N_1}{N_2} = a = \frac{I_{b2}}{I_{b1}} \tag{1.38}$$

From (1.10) and (1.38), we have:

$$\hat{V}_1^{\text{pu}} = \frac{\hat{V}_1^V}{V_{b1}^V} = \frac{(a)\hat{V}_2^V}{(a)V_{b2}^V} = \hat{V}_2^{\text{pu}}, \quad \hat{I}_1^{\text{pu}} = \frac{\hat{I}_1^A}{I_{b1}^A} = \frac{(1/a)\hat{I}_2^A}{(1/a)I_{b2}^A} = \hat{I}_2^{\text{pu}} \tag{1.39}$$

1.7.4 The Relationship between Nominal, Base, and Perunit

Usually, the base of each element is the nominal of that element.

$$S_b^{VA} = S_n^{VA}, \quad V_b^V = V_n^V, \quad I_b^A = I_n^A \tag{1.40}$$

In other words, we have a nominal load in perunit:

$$\Rightarrow I_n^{\text{pu}} = \frac{I_n^A}{I_b^A} = \frac{I_n^A}{I_n^A} = 1^{\text{pu}} \Rightarrow V_n^{\text{pu}} = 1^{\text{pu}} \Rightarrow S_n^{\text{pu}} = 1^{\text{pu}} \tag{1.41}$$

We have a nominal half-load in perunit:

$$\Rightarrow I_{0.5n}^{\text{pu}} = 0.5^{\text{pu}}, \quad V_{0.5n}^{\text{pu}} = 1^{\text{pu}}, \quad S_{0.5n}^{\text{pu}} = 0.5^{\text{pu}} \tag{1.42}$$

1.8 CORE LOSS (HYSTERESIS AND EDDY CURRENT)

The core losses, also known as iron losses, no-load losses, or open-circuit losses, are the same. These losses are the sum of hysteresis and eddy current losses.

The hysteresis loss is:

$$P_h = k_h(f)(\phi_m)^n, \quad n = 1.5:2.5, \quad (\text{Usually} = 2) \tag{1.43}$$

The eddy current loss is:

$$P_e = k_e\left(f^2\right)\left(\phi_m\right)^2 \tag{1.44}$$

By approximating ($\phi \propto \frac{V}{f}$), the equations (1.43) and (1.44) can be written as follows:

$$P_h = k_h\frac{V^2}{f}, \quad P_e = k_e V^2, \quad \Rightarrow P_c = P_h + P_e = k_h\frac{V^2}{f} + k_e V^2 \tag{1.45}$$

Note that core loss is a function of voltage and frequency, and not a function of load and current. Meaning that the core loss is constant from no-load to full-load condition.

1.9 EFFICIENCY AND MAXIMUM EFFICIENCY

The transformer efficiency is defined as follows:

$$\%\eta = 100\frac{P_{\text{out}}}{P_{\text{in}}} = 100\frac{P_{\text{out}}}{P_{\text{out}} + P_{\text{loss}}} = 100\frac{P_{\text{out}}}{P_{\text{out}} + P_{\text{cu}} + P_c} \tag{1.46}$$

Daily or annual (day or year) efficiency is defined as follows:

$$\%\eta = 100\frac{P_{\text{out}}(\text{day, year})}{P_{\text{in}}(\text{day, year})} = 100\frac{P_{\text{out}}(\text{day, year})}{P_{\text{out}}(\text{day, year}) + P_{cu}(\text{day, year}) + P_c(\text{day, year})} \tag{1.47}$$

Maximum efficiency occurs under the following conditions:

$$\text{Constant load-independent loss}\left(\text{core loss}\right) = \text{load-dependent loss}\left(\text{copper loss}\right) \tag{1.48}$$

$$\eta_{\max} \Leftrightarrow P_c = P_{\text{cu}} \tag{1.49}$$

In the approximate models (b) and (c) of Figure 1.10, copper losses can be written as follows:

$$P_{\text{cu}} = R_{\text{eq}} I_2'^2 \tag{1.50}$$

Note: if the load is halved, the copper loss will be reduced to one-quarter.

1.10 OPEN- AND SHORT-CIRCUIT TEST OF A TRANSFORMER

These two tests are used to determine the parameters and performance characteristics of the transformer.

1.10.1 Open-Circuit or No-Load Test (O.C)

In this case, the transformer is no-load and the primary is connected to the rated voltage. The voltage ($V_{\text{o.c}}$), current ($I_{\text{o.c}}$), and input power ($P_{\text{o.c}}$) of the transformer are measured. Using the equivalent circuit (b) in Figure 1.10, the series branch is removed and we have:

$$\varphi_{\text{o.c}} = \cos^{-1}\left(\frac{P_{\text{o.c}}}{V_{\text{o.c}} I_{\text{o.c}}}\right) \Rightarrow R_c = \frac{V_{\text{o.c}}}{I_{\text{o.c}}\cos\varphi_{\text{o.c}}}, \quad X_m = \frac{V_{\text{o.c}}}{I_{\text{o.c}}\sin\varphi_{\text{o.c}}} \tag{1.51}$$

Also:

$$V_{o.c} = V_n = 1\,\text{pu}, \quad P_{o.c} = P_c \tag{1.52}$$

1.10.2 Short-Circuit Test (S.C)

In this case, the secondary is short-circuited and the primary voltage is much lower than the rated voltage so that the input current is rated. In this case, the voltage ($V_{s.c}$), current ($I_{s.c}$), and input power ($P_{s.c}$) of the transformer are measured.

Using the equivalent circuit (a) in Figure 1.10, the parallel branch is removed and we have:

$$R_{eq} = \frac{P_{s.c}}{I_{s.c}^2}, \quad |Z_{eq}| = \frac{V_{s.c}}{I_{s.c}}, \quad X_{eq} = \sqrt{|Z_{eq}|^2 - R_{eq}^2}, \quad Z_{eq} = R_{eq} + jX_{eq} \tag{1.53}$$

Then:

$$R_1 = R_2' = 0.5R_{eq}, \quad X_1 = X_2' = 0.5X_{eq} \tag{1.54}$$

We have:

$$I_{s.c} = I_n = 1^{pu}, \quad P_{s.c} = P_{cun} \tag{1.55}$$

1.11 CONNECTIONS OF TRANSFORMERS

Two transformers can be connected in 4 ways as shown in Figure 1.12.

Only the parallel type of connection is practical in power systems (Figure 1.12 (d)). Usually, only the simple model (Figure 1.10c and d) is used for simple calculations (See Figure 1.13).

The equations of two parallel transformers can be written as follows:

$$S = \hat{V}\hat{I}^*, \quad S_1 = \hat{V}\hat{I}_1^*, \quad S_2 = \hat{V}\hat{I}_2^*, \quad S = S_1 + S_2 \tag{1.56}$$

$$\hat{I}_1 = \frac{Z_2}{Z_1 + Z_2}\hat{I}, \quad \hat{I}_2 = \frac{Z_1}{Z_1 + Z_2}\hat{I}, \quad \frac{\hat{I}_1}{\hat{I}_2} = \frac{Z_2}{Z_1}, \quad \frac{S_1}{S_2} = \frac{\hat{I}_1^*}{\hat{I}_2^*} \tag{1.57}$$

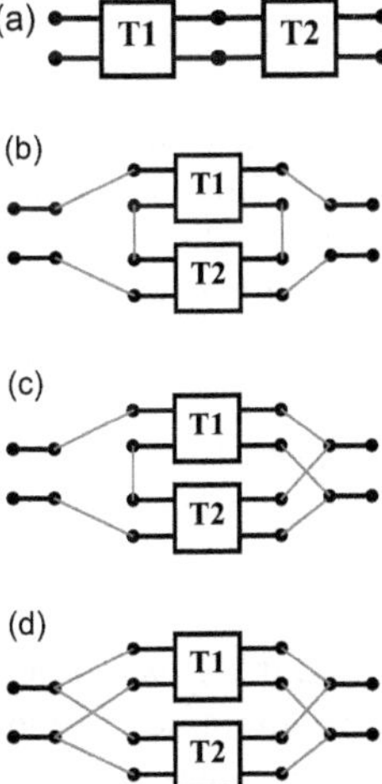

FIGURE 1.12 Types of connecting two transformers: (a) series (or cascade), (b) series–series, (c) series–parallel, (d) parallel (or parallel–parallel).

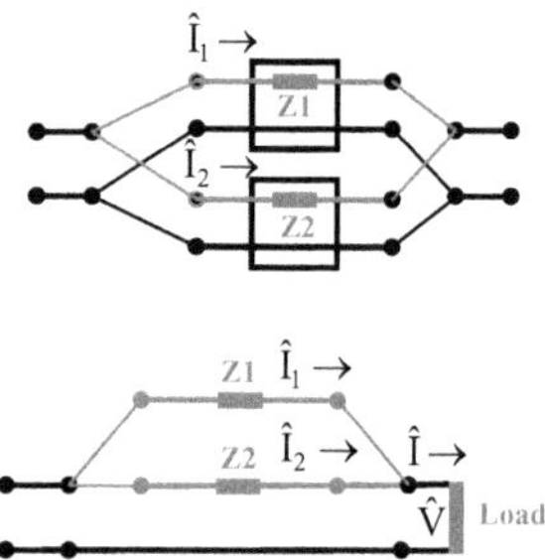

FIGURE 1.13 Parallel type of connection between two transformers.

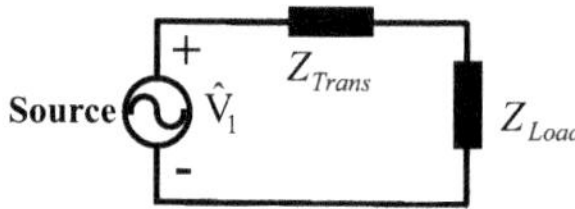

FIGURE 1.14 Maximum power transfer.

1.12 MAXIMUM POWER TRANSFER THEOREM

Look at Figure 1.14. To transfer the maximum power from the voltage source to the load, the relation (1.58) must be satisfied.

$$Z_{\text{trans}} = Z^*_{\text{load}} \Rightarrow R_{\text{trans}} + jX_{\text{trans}} = R_{\text{load}} - jX_{\text{load}} \tag{1.58}$$

Part Two: Answer-Question

1.13 TWO-CHOICE QUESTIONS (YES/NO)

1. A high-power transformer can be assumed to be ideal.
2. In no-load condition, a real transformer behaves similar to a large inductor.
3. The smaller the load power factor, the lower the efficiency.
4. In the transformer no-load test, copper losses are measured.
5. No-load losses, open circuit losses, and core losses are equal.
6. The efficiency of a transformer depends on the load.
7. The resistance between the primary and secondary windings of the transformer is zero.
8. An ideal transformer transfers power without losses.
9. The ideal transformer is removed due to the way the bases are selected.
10. With a step-up transformer, you can only make a step-up autotransformer.
11. The main issue with the autotransformer is the electrical connection between its primary and secondary windings.
12. The transformer's inrush current can be as high as ten times the rated current.
13. The power transformers are usually air-cooled.
14. The load is divided between two parallel transformers, inversely proportional to the impedance.
15. Power transformers are usually rated in terms of kilowatts.
16. Voltage regulation is always positive in phase-lag and resistive loads.
17. Voltage regulation is always negative at phase-lead load.
18. The reason for using an autotransformer is to reduce losses.
19. Generally, core loss increases with increasing frequency.
20. If the working frequency of a transformer is reduced, the voltage should also be reduced in the same ratio.

21. If an autotransformer is made from a transformer, the power ratio of the autotransformer to the transformer is always greater than one.
22. The no-load current is five times the rated current.
23. The main reason for using a magnetic core in transformers is to reduce losses.
24. Usually, the side of the high voltage level of the transformer is short-circuited in the short-circuit test.
25. Voltage and the number of turns of the coil have a direct relationship.
26. If the load changes from 0.8 lagging to 0.8 leading, the efficiency changes.
27. Taps (Tap Changers) are provided at the HV windings of the transformer.
28. The humming sound produced by transformers is due to electromagnetic forces acting on the transformer's core.
29. A Buchholz relay is a safety device mounted on some oil-filled power transformers.
30. A Buchholz relay is used instead of a "gas detector relay".
31. The transformer core is sheeted to reduce eddy current losses.
32. The transformer core is laminated to reduce hysteresis losses.
33. The silicon steel used as a transformer core has less hysteresis.
34. Sumpner's test is used to determine temperature rise, efficiency, and voltage regulation (this is not included in the text).
35. The efficiencies of power transformers typically vary from 97% to 99%.
36. Weight, size, and cost of the transformer are less with a cruciform core.

1.14 KEY ANSWERS TO TWO-CHOICE QUESTIONS

Yes	1,2,5,6,8,9,11,12,14,16,19,20,21,25,27,28,29,31,33,34,35,36
No	3,4,7,10,13,15,17,18,22,23,24,26,30,32

1.15 DESCRIPTIVE QUESTIONS OF THE SINGLE-PHASE TRANSFORMERS

1.1 Find the currents in Figure 1.15.

(Difficulty level ○ Easy ● Normal ○ Hard)

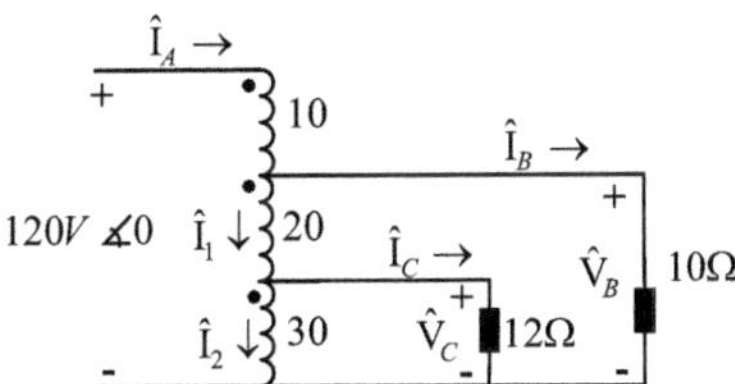

FIGURE 1.15 Question Figure 1.1.

1.2 Find the currents in Figure 1.16.

(Difficulty level ○ Easy ● Normal ○ Hard)

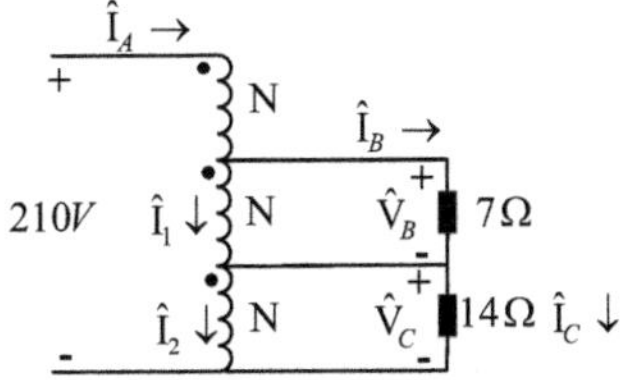

FIGURE 1.16 Question Figure 1.2.

1.3 Find the currents in Figure 1.17.

(Difficulty level ○ Easy ● Normal ○ Hard)

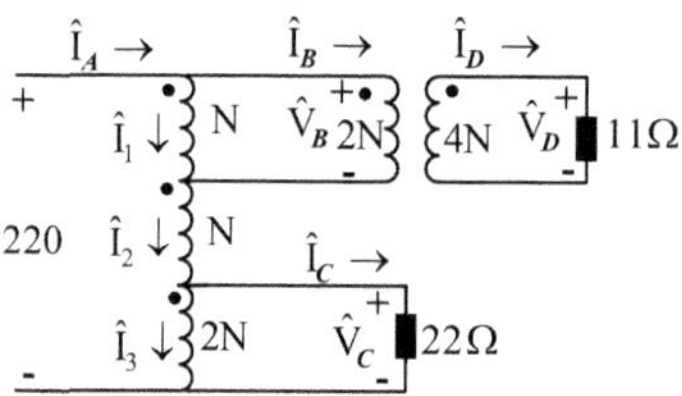

FIGURE 1.17 Question Figure 1.3.

1.4 Find the voltages in Figure 1.18.

(Difficulty level ● Easy ○ Normal ○ Hard)

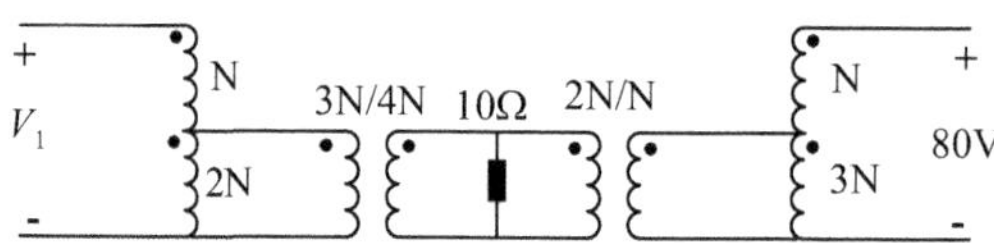

FIGURE 1.18 Question Figure 1.4.

1.5 Find the (I) in Figure 1.19.

(Difficulty level ● Easy ○ Normal ○ Hard)

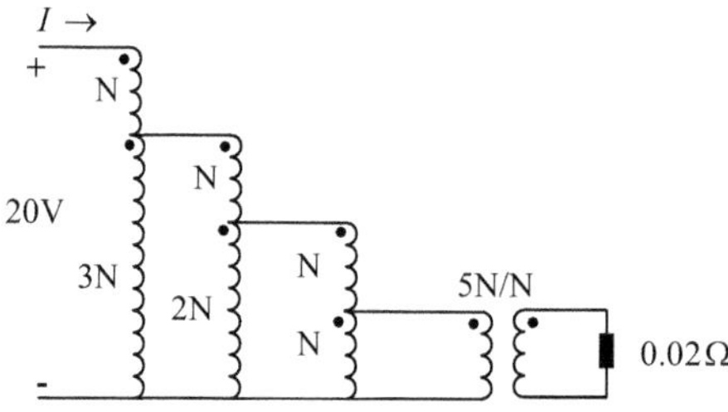

FIGURE 1.19 Question Figure 1.5.

1.6 Find the (I_D) in Figure 1.20.

(Difficulty level ● Easy ○ Normal ○ Hard)

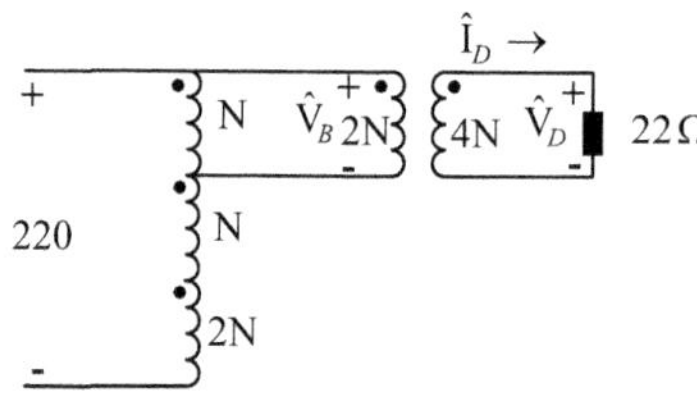

FIGURE 1.20 Question Figure 1.6.

1.7 (See Figure 1.21) Find (Z_B/Z_C) such that ($I_{1=0}$).

(Difficulty level ○ Easy ● Normal ○ Hard)

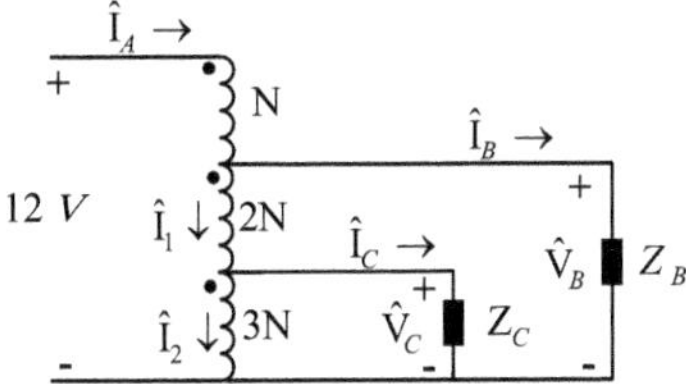

FIGURE 1.21 Question Figure 1.7.

1.8 Find the (P_{in}) in Figure 1.22.

(Difficulty level ○ Easy ● Normal ○ Hard)

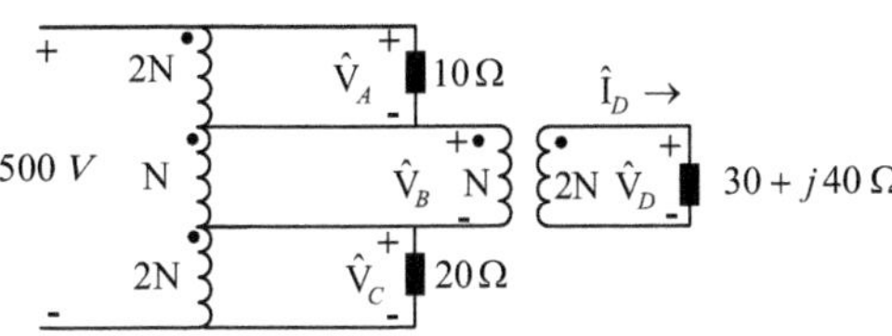

FIGURE 1.22 Question Figure 1.8.

1.9 Consider a 100 kVA, 10/2 kV, 60 Hz transformer. Assume:

$$R_1 = 6.1\,\Omega,\quad R_2' = 7.2\,\Omega,\quad X_1 = X_2' = 31.2\,\Omega,\quad X_m = 57\text{ k}\Omega,\quad R_c = 12\text{ k}\Omega$$

$$\text{Load}: 100\,\Omega,\quad \text{PF} = 0.8\text{ lead, } 2\text{ kV}$$

Find V_1, I_1 and η.

(Difficulty level ○ Easy ● Normal ○ Hard)

1.10 Consider a 10 kVA, 2,400/240 V, 60 Hz transformer. Assume the load is nominal and $R_1 = R_2' = 5.2\,\Omega$, $X_1 = X_2' = 7.6\,\Omega$

Find voltage regulation if (a) PF=0.8 lag, (b) PF=0.8 lead.

(Difficulty level ● Easy ○ Normal ○ Hard)

1.11 Consider a transformer with 2% resistance and 10% reactance. Find voltage regulation and efficiency if we have half of the nominal load and PF=0.7 lead.

(Difficulty level ● Easy ○ Normal ○ Hard)

1.12 The maximum efficiency of the transformer is 80% and occurs at full load. The transformer works for 16 hours at full load and 8 hours at half-full load. Find the daily efficiency of the transformer.

(Difficulty level ● Easy ○ Normal ○ Hard)

1.13 If two parallel transformers feed a 4 kVA load at a unit power factor, what power does each transformer transfer? If: $Z_{T1} = 100 + j50\,\Omega$, $Z_{T2} = 110 + j70\,\Omega$

(Difficulty level ● Easy ○ Normal ○ Hard)

1.14 Consider 11 kVA, 1 kV/400 V transformer. Assume the data of the short-circuit test are obtained as follows: (30 V, 10 A, 100 W). Find the load voltage of the transformer if the transformer receives its rated load at the unit power factor from the source.

(Difficulty level ● Easy ○ Normal ○ Hard)

1.15 Assume the data of the open-circuit test are obtained as follows: (400 V, 5 A, 500 W). Find the core-loss and magnetizing component of the excitation current.

(Difficulty level ● Easy ○ Normal ○ Hard)

1.16 Consider 10 kVA, 1 kV/500 V transformer. Assume the data of the short-circuit test are obtained as follows: (25 V, 10 A, 90 W). Find the total transformer impedance (Z_{eq}).

1.17 Consider 100 kVA, 10 kV/500 V transformer. Assume the data of short- and open-circuit tests are obtained as follows: (100 V, 10 A, 1,000 W), (10 kV, 2 A, 500 W). Find the efficiency of the transformer at one-third rated load and 0.7 lead.

(Difficulty level ○ Easy ● Normal ○ Hard)

1.18 The results of two open-circuit tests on a transformer are as follows:

$$(V_1, f_1, P_1), (V_2, f_2, P_2)$$

Find the relationship between core losses in terms of voltage and frequency.

(Difficulty level ● Easy ○ Normal ○ Hard)

1.19 The results of two short-circuit tests on a transformer are as follows:

(100 V, 10 A, 50 Hz), (100 V, 1.15 A, 500 Hz)

Find the total transformer impedance (Z_{eq}) at the frequency of 50 Hz.

(Difficulty level ● Easy ○ Normal ○ Hard)

1.20 Find the core losses of a 500 kVA transformer under the following conditions.

$$(1)\ \%\eta = 98, \text{Full Load}, 0.8\ \text{lag}, \quad (2)\ \%\eta = 99, \text{Half-Full Load}, \text{PF} = 1$$

(Difficulty level ● Easy ○ Normal ○ Hard)

1.16 DESCRIPTIVE ANSWERS OF THE SINGLE-PHASE TRANSFORMERS

1.1 (See Figure 1.23) There exists a direct relationship between voltage and coil turns. Thus, the following formula is used to determine the voltage at each coil turn.

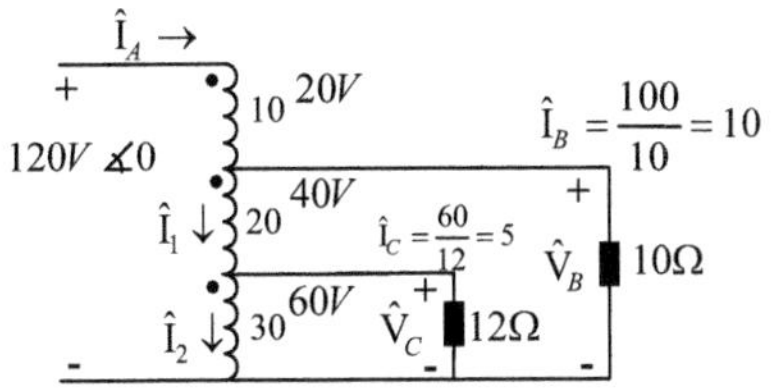

FIGURE 1.23 Answer Figure 1.1.

$$\frac{120\text{ V}}{10+20+30}=2\left(\frac{\text{V}}{\text{Turn}}\right)\Rightarrow\begin{cases}V_B=50(2)=100\text{ V}\\ V_C=30(2)=60\text{ V}\end{cases}\Rightarrow I_B=\frac{100}{10}=10\text{ A},\quad I_C=\frac{60}{12}=5\text{ A}$$

We have:

$$S_{\text{in}}=S_{\text{out}}\Rightarrow 120(I_A)=100(10)+60(5)\Rightarrow I_A=10.833\text{ A}$$

$$\text{KCL}\Rightarrow I_1=I_A-I_B=10.833-10=0.833\text{ A}$$

$$\text{KCL}\Rightarrow I_2=I_1-I_C=0.833-5=-4.167\text{ A}$$

We can now verify the computations. From equation (1.9):

$$\sum_{i=1}^{3}\left(N_i\hat{I}_i\right)=0\Rightarrow 10(I_A)+20(I_1)+30(I_2)=10(10.833)+20(0.833)+30(-4.167)=-0.02$$

$$\Rightarrow\sum_{i=1}^{3}\left(N_i\hat{I}_i\right)=-0.02\simeq 0$$

1.2 (See Figure 1.24) Similar to question 1.1.

$$\frac{210\text{ V}}{N+N+N}=70\left(\frac{\text{V}}{\text{Turn}}\right)\Rightarrow V_B=V_C=70\text{ V}\Rightarrow I_B=\frac{70}{7}=10\text{ A},\quad I_C=\frac{70}{14}=5\text{ A}$$

$$S_{\text{in}}=S_{\text{out}}\Rightarrow 210(I_A)=70(10)+70(5)\Rightarrow I_A=5\text{ A}$$

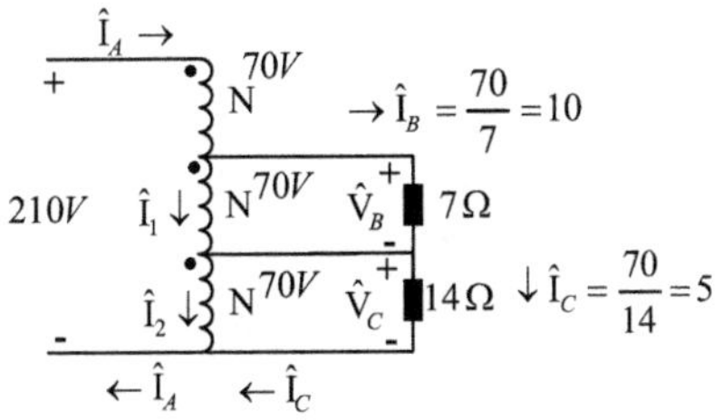

FIGURE 1.24 Answer Figure 1.2.

$$\text{KCL} \Rightarrow I_1 = I_A - I_B = 5 - 10 = -5 \text{ A}$$

$$\text{KCL} \Rightarrow I_2 = I_A - I_C = 5 - 5 = 0 \text{ A}$$

We can now verify the computations. From equation (1.9):

$$\sum_{i=1}^{3}\left(N_i \hat{\text{I}}_i\right) = 0 \Rightarrow N(I_A) + N(I_1) + N(I_2) = N(5) + N(-5) + N(0) = 0$$

1.3 Similar to question 1.1, see Figures 1.25 and 1.26.

$$\frac{220 \text{ V}}{N + N + 2N} = 55\left(\frac{V}{\text{Turn}}\right) \Rightarrow V_B = 55 \text{ V}, \quad V_C = 110 \text{ V} \Rightarrow I_C = \frac{110}{22} = 5 \text{ A}$$

$$\Rightarrow \frac{V_D}{V_B} = \frac{4\ N}{2\ N} \Rightarrow V_D = 2(55) = 110 \text{ V} \Rightarrow I_D = \frac{110}{11} = 10 \text{ A}$$

$$\Rightarrow \frac{I_B}{I_D} = \frac{4\ N}{2\ N} \Rightarrow I_B = 2(10) = 20 \text{ A}$$

$$S_{\text{in}} = S_{\text{out}} \Rightarrow 220(I_A) = V_D(I_D) + V_C(I_C) = 110(10) + 110(5) \Rightarrow I_A = 7.5 \text{ A}$$

$$\text{KCL} \Rightarrow I_1 = I_A - I_B = 7.5 - 20 = -12.5 \text{ A}, \quad \text{KCL} \Rightarrow I_2 = I_1 + I_B = -12.5 + 20 = 7.5 \text{ A}$$

$$\text{KCL} \Rightarrow I_3 = I_2 - I_c = 7.5 - 5 = 2.5 \text{ A}$$

We can now verify the computations. From equation (1.9):

$$\sum_{i=1}^{3}\left(N_i \hat{I}_i\right) = 0 \Rightarrow N(-12.5) + N(7.5) + 2N(2.5) = 0$$

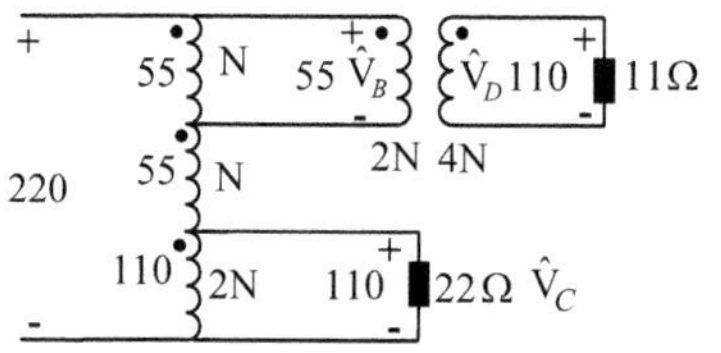

FIGURE 1.25 Answer Figure 1.3 (voltage).

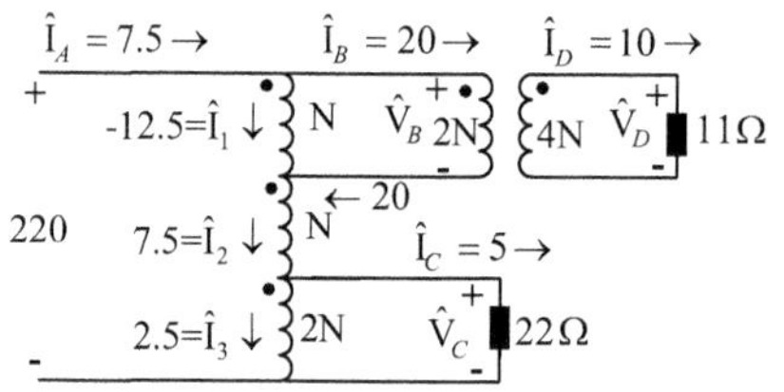

FIGURE 1.26 Answer Figure 1.3 (current).

1.4 (See Figure 1.27) Using the ratio of the coil turns, we compute the voltage from the right to the left.

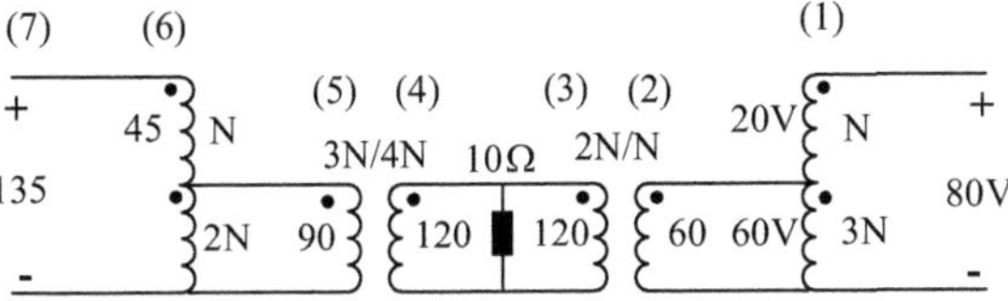

FIGURE 1.27 Answer Figure 1.4 (voltage).

1.5 (See Figure 1.28) Using the ratio of the coil turns, we compute the voltages from the left to the right.

$$S_{in} = S_{out} \Rightarrow 20(I) = \frac{V^2}{R} = \frac{1}{0.02} = 50 \Rightarrow I = 2.5 \text{ A}$$

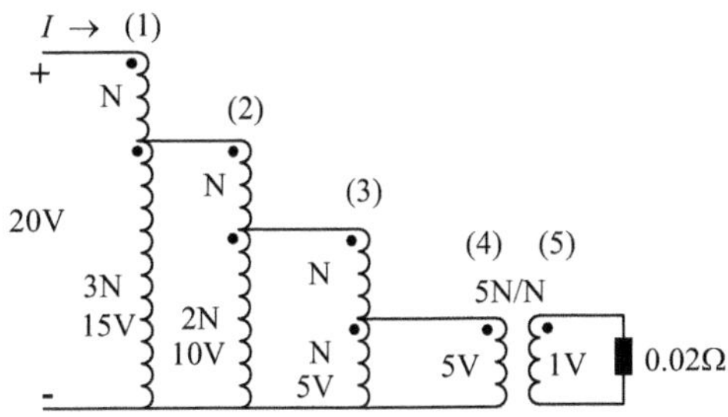

FIGURE 1.28 Answer Figure 1.5 (voltage).

1.6 (See Figure 1.29). Using the ratio of the coil turns, we compute the voltage from the left to the right.

$\hat{I}_D = \frac{110}{22} = 5A$ (4)

(1)
+
55 V
N
$\hat{V}_B$ 2N
4N
$\hat{V}_D$
22Ω
220V
N
55 V
110 V
(2)
(3)
2N
-

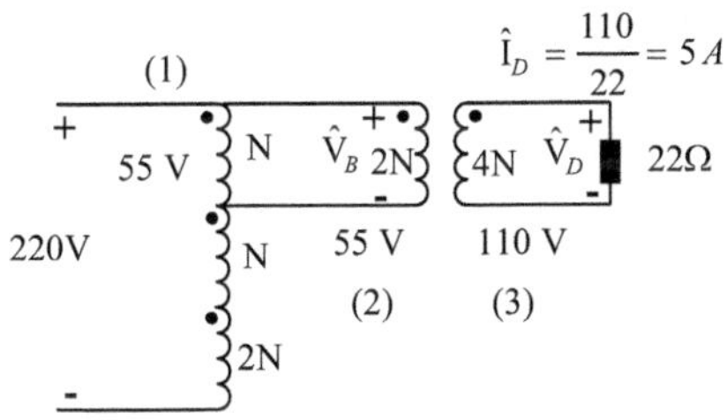

FIGURE 1.29 Answer Figure 1.6 (voltage).

1.7 See Figure 1.30.

From equations (1.8) and (1.9):

$$V_B = \frac{5N}{6N}(12 \text{ V}) = 10 \text{ V}, \quad V_C = \frac{3N}{6N}(12 \text{ V}) = 6 \text{ V}$$

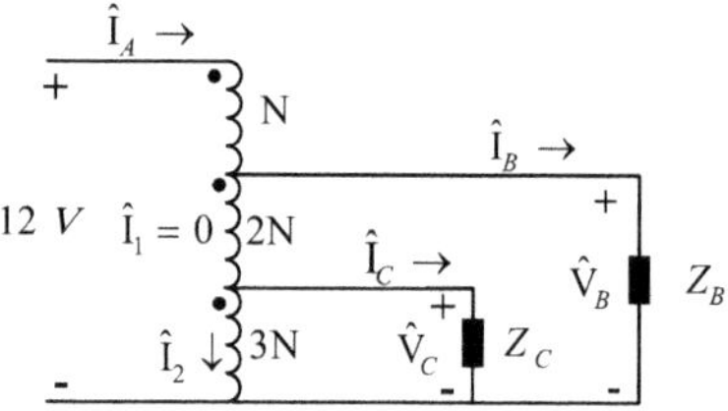

FIGURE 1.30 Answer Figure 1.7.

$$\text{KVL}: V_B = 10 = Z_B I_B, \quad V_C = 6 = Z_C I_C \Rightarrow \frac{Z_B}{Z_C}\frac{I_B}{I_C} = \frac{10}{6} \tag{1.59}$$

$$\text{KCL}: I_1 = 0 \Rightarrow I_A = I_B, \quad I_2 = -I_C \tag{1.60}$$

$$\text{From (1.9): } N(I_A) + 2N(0) + 3N(I_2) = 0 \Rightarrow \frac{I_A}{I_2} = -3 \overset{(1.60)}{\Rightarrow} \frac{I_B}{I_C} = 3 \tag{1.61}$$

$$\text{From (1.59) and (1.61): } \frac{Z_B}{Z_C}(3) = \frac{10}{6} \Rightarrow \frac{Z_B}{Z_C} = \frac{5}{9}$$

1.8 (See Figure 1.31) Using the ratio of the coil turns, we compute the voltage from the left to the right.

$$\text{Load side KVL: } I_D = \frac{200}{\sqrt{30^2 + 40^2}} = 4 \text{ A}$$

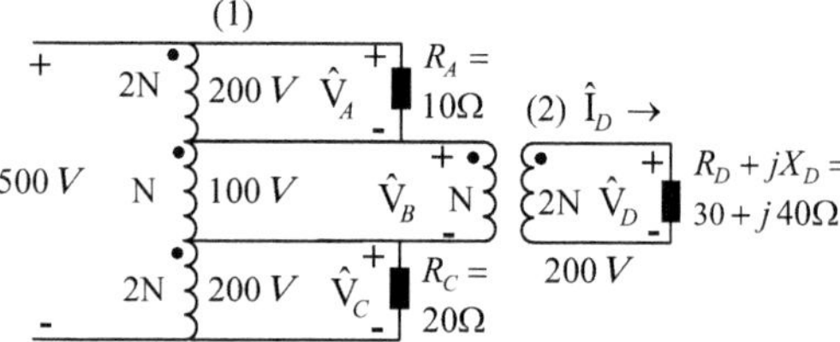

FIGURE 1.31 Answer Figure 1.8.

$$P_{in} = P_{out} \Rightarrow P_{in} = \frac{V_A^2}{R_A} + \frac{V_C^2}{R_C} + R_D I_D^2 = \frac{200^2}{10} + \frac{200^2}{20} + 30(4)^2 = 6480 \text{ W}$$

1.9 See Figure 1.32.

$$I_2 = \frac{V_2}{|Z|} = \frac{2 \text{ kV}}{100\,\Omega} = 20 \text{ A} \Rightarrow \hat{I}_2 = 20\measuredangle\cos^{-1}(0.8), \ \hat{V}_2 = 2 \text{ kV}\measuredangle 0$$

$$\Rightarrow \hat{V}_2' = \frac{10}{2}(2 \text{ kV}) = 10\,\text{kV}\measuredangle 0, \ \hat{I}_2' = \frac{2}{10}\left(20\measuredangle\cos^{-1}(0.8)\right) = 4\measuredangle\cos^{-1}(0.8)$$

$$\text{KVL}: \hat{E} = (7.2 + j31.2)(4\measuredangle\cos^{-1}(0.8)) + 10\,k\measuredangle 0 = 9949\measuredangle 0.675°$$

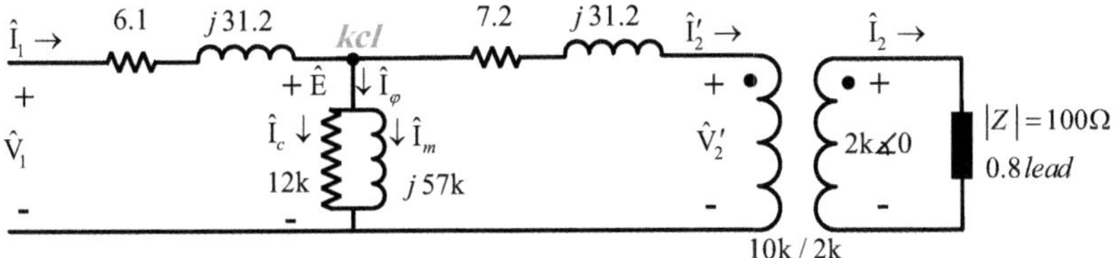

FIGURE 1.32 Answer Figure 1.9.

$$\text{KVL}: \begin{cases} \hat{I}_c = \dfrac{\hat{E}}{R_c} = \dfrac{9949\measuredangle 0.675}{12\,\text{k}} = 0.829\measuredangle 0.675 \\ \hat{I}_m = \dfrac{\hat{E}}{jX_m} = \dfrac{9949\measuredangle 0.675}{j57\,\text{k}} = 0.175\measuredangle -89.33 \end{cases}$$

$$\text{KCL}: \ \hat{I}_1 = \hat{I}_c + \hat{I}_m + \hat{I}'_2 = 4.609\measuredangle 29^\circ$$

$$\text{KVL}: \hat{V}_1 = (6.1 + j31.2)(4.609\measuredangle 29) + 9949\measuredangle 0.675 = 9907\measuredangle 1.484^\circ$$

$$\Rightarrow \cos\varphi_1 = \cos(1.484 - 29) = 0.887$$

$$P_{\text{loss}} = P_{\text{cu}} + P_c = (6.1)(4.609^2) + (7.2)(4^2) + (12\,\text{k})(0.829^2) = 8492\,\text{W}$$

$$P_{\text{out}} = V_2 I_2 \cos\varphi_2 = (2\,\text{k})(20)(0.8) = 32\,\text{kW}$$

$$\%\eta = \frac{32\,\text{k}}{32\,\text{k} + 8492}(100) = 79.03\%$$

Checking the law of conservation of energy:

$$P_{\text{in}} = V_1 I_1 \cos\varphi_1 = (9907)(4.609)(0.887) = 40.502 \ \text{kW}$$

$$P_{\text{out}} + P_{\text{loss}} = 32\,\text{k} + 8492 = 40.492\,\text{kW} \simeq P_{\text{in}}$$

1.10 In this case, we use the approximate model of Figure 1.10c. We have:

$$R_{\text{eq}} = R_1 + R'_2 = 10.4, \quad X_{\text{eq}} = X_1 + X'_2 = 15.2, \text{ See Figure 1.33.}$$

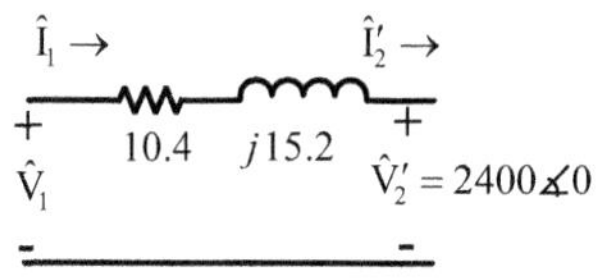

$\hat{I}_1 \rightarrow$ $\hat{I}_2' \rightarrow$

$+$ 10.4 $j15.2$ $+$

$\hat{V}_1$ $\hat{V}_2' = 2400\measuredangle 0$

$-$ $-$

FIGURE 1.33 Answer Figure 1.10.

From Section 1.5, we have: $I_{1n} = \dfrac{S_n}{V_{1n}} = \dfrac{10\,\text{kVA}}{2400\,\text{V}} = 4.167\,\text{A}$

(a) $\Rightarrow \hat{I}_2' = 4.167\measuredangle - \cos^{-1} 0.8$

$$\hat{V}_1 = (10.4 + j15.2)(4.167\measuredangle - \cos^{-1} 0.8) + 2400\measuredangle 0 = 2473\measuredangle 0.572^\circ$$

From equation (1.33), we have: $\%\ V_R = (100)\dfrac{2473 - 2400}{2400} = 3.04\%$

The second method (the approximate method of equation (1.34)):

$$P = VI\cos\varphi = (2400)(4.167)(0.8) = 8000.64 \simeq 8000 = |S|\cos\varphi = 10\text{k}(0.8)$$

$$Q = VI\sin\varphi = (2400)(4.167)(0.6) = 6000.48 \simeq 6000 = |S|\sin\varphi = 10\text{k}(0.6)$$

$$\Rightarrow \%\ V_R = (100)\frac{(10.4)(8000) + (15.2)(6000)}{(2400)^2} = 3.03\%$$

(b) $\Rightarrow \hat{I}_2' = 4.167\measuredangle + \cos^{-1} 0.8$

$$\hat{V}_1 = (10.4 + j15.2)(4.167\measuredangle + \cos^{-1} 0.8) + 2400\measuredangle 0 = 2398\measuredangle 1.83^\circ$$

From equation (1.33), we have: $\%\ V_R = (100)\dfrac{2398 - 2400}{2400} = -0.083\% \simeq -0.1\%$

The second method (the approximate method of equation (1.34)):

$$\Rightarrow \%\ V_R = (100)\frac{(10.4)(8000) - (15.2)(6000)}{(2400)^2} = -0.139\% \simeq -0.1\%$$

1.11 When the parameters are expressed in terms of percentage, it means that the system is perunited. From Section 1.7.4 we have:

$$R_{eq} = 0.02\ \text{pu},\ X_{eq} = 0.1\ \text{pu (See Figure 1.34)}$$

Half of the nominal load and PF=0.7 lead in perunit means $\hat{V}_2' = 1,\ \hat{I}_2' = 0.5\measuredangle \cos^{-1} 0.7$

$$\hat{V}_1 = (0.02 + j0.1)(0.5\measuredangle \cos^{-1} 0.7) + 1\measuredangle 0 = 0.972\measuredangle 2.484^\circ$$

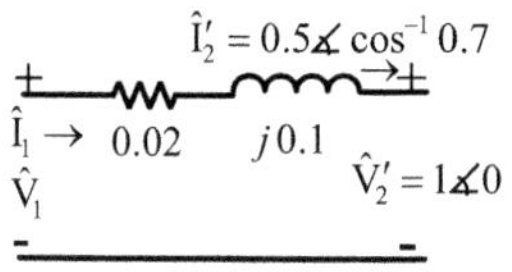

$\hat{I}_2' = 0.5\measuredangle \cos^{-1} 0.7$

$+$ $\rightarrow +$

$\hat{I}_1 \rightarrow$ 0.02 $j0.1$

$\hat{V}_1$ $\hat{V}_2' = 1\measuredangle 0$

$-$ $-$

FIGURE 1.34 Answer Figure 1.11.

From equation (1.33), we have: $\% \ V_R = (100)\dfrac{0.972-1}{1} = -2.8\%$

The second method (the approximate method of equation (1.34)):

$$P = VI\cos\varphi = (1)(0.5)(0.7) = 0.35\,\text{pu}$$

$$Q = VI\sin\varphi = (1)(0.5)\left(\sin\left(\cos^{-1}0.7\right)\right) = 0.357\,\text{pu}$$

On the other hand: $Q = \sqrt{|S|^2 - P^2} = \sqrt{0.5^2 - 0.35^2} = 0.357\,\text{pu}$

$$\Rightarrow \% \ V_R = (100)\frac{(0.02)(0.35)-(0.1)(0.357)}{(1)^2} = -2.87\%$$

Efficiency: $P_{\text{loss}} = R_{\text{eq}}I^2 = 0.02(0.5)^2 = 0.005$, $P_{\text{out}} = VI\cos\varphi = (1)(0.5)(0.7) = 0.35\,\text{pu}$

$$P_{\text{in}} = \begin{cases} = V_1I_1\cos\varphi_1 = (0.972)(0.5)\cos(2.484 - \cos^{-1}0.7) = 0.355 \\ = P_{\text{out}} + P_{\text{loss}} = 0.35 + 0.005 = 0.355 \end{cases}$$

$$\Rightarrow \%\eta = 100\frac{P_{\text{out}}}{P_{\text{in}}} = 100\frac{0.35}{0.355} = 98.59\%$$

1.12 From equations (1.49) and (1.50), we have:

The maximum efficiency of the transformer occurs at full load, then: $P_c = P_{\text{cun}}$, and:

$$\eta_{\max} = \frac{P_{\text{out}}}{P_{\text{out}} + P_c + P_{\text{cun}}} \Rightarrow 0.8 = \frac{1}{1+2P_c} \Rightarrow P_c = P_{\text{cun}} = 0.125$$

(1): At full load: $P_{\text{out}} = 1,\ P_c = P_{\text{cun}} = 0.125$

(2): At half-full load: $P_{\text{out}} = 0.5,\ P_c = 0.125,\ P_{\text{cu}} = \dfrac{0.125}{2^2} = 0.03125$

From equation (1.47), we have:

$$\%\eta = \frac{\overbrace{(16)}^{\text{hour}}\overbrace{(1)}^{P_{\text{out}}} + \overbrace{(8)}^{\text{hour}}\overbrace{(0.5)}^{P_{\text{out}}}}{\underbrace{(16)}_{\text{hour}}(\underbrace{1}_{P_{\text{out}}} + \underbrace{0.125}_{P_c} + \underbrace{0.125}_{P_{\text{cun}}}) + \underbrace{(8)}_{\text{hour}}(\underbrace{0.5}_{P_{\text{out}}} + \underbrace{0.125}_{P_c} + \underbrace{0.03125}_{P_{\text{cu}}})}(100) = 79.2\%$$

1.13 $PF = \cos\varphi = 1 \Rightarrow \varphi = 0,\ |S| = P,\ Q = 0 \Rightarrow P_1 + P_2 = 4\ \text{kW}\ \ (1)$

From equation (1.57), we have: $\dfrac{P_1}{P_2} = \dfrac{|Z_2|}{|Z_1|} = \dfrac{\sqrt{110^2+70^2}}{\sqrt{100^2+50^2}} = 1.166\ \ (2)$

$$\overset{(1),(2)}{\Rightarrow}\ 1.166P_2 + P_2 = 4\ \text{kW} \Rightarrow P_2 = 1.847\ \text{kW},\ P_1 = 2.153\ \text{kW}$$

1.14 See Figure 1.35.

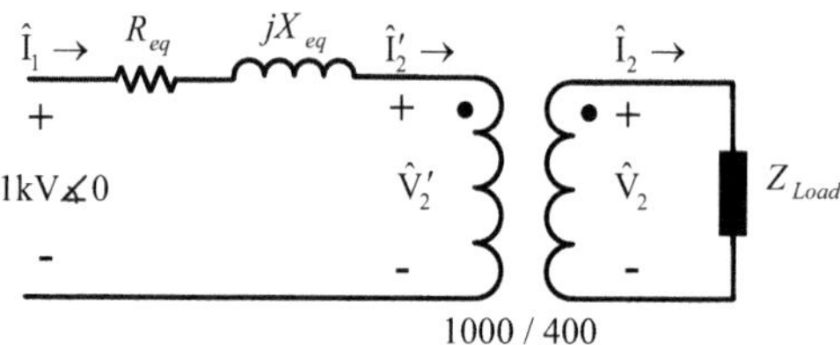

FIGURE 1.35 Answer Figure 1.14.

From equation (1.53), we have: $R_{eq} = \frac{100}{10^2} = 1,\ |Z_{eq}| = \frac{30}{10} = 3,\ X_{eq} = \sqrt{3^2 - 1} = \sqrt{8}$

Transformer receives its rated load at unit power factor from the source, then:

$$I_1 = I_{n1} = \frac{S_n}{V_{n1}} = \frac{11\,\text{kVA}}{1\,\text{kV}} = 11\,\text{A} \Rightarrow \hat{I}_1 = \hat{I}_2' = 11\,\text{A}\measuredangle 0$$

$$\text{KVL: } \hat{V}_2' = 1000\measuredangle 0 - (1 + j\sqrt{8})(11\measuredangle 0) = 989.49\measuredangle -1.802^\circ$$

$$\Rightarrow V_2 = \frac{400}{1000}(989.49) = 395.8\,\text{V}$$

1.15 From equation (1.51), we have:

$$\cos\varphi_{o.c} = \frac{P_{o.c}}{(V_{o.c})(I_{o.c})} = \frac{500}{(400)(5)} = 0.25 \Rightarrow \sin\varphi_{o.c} = \sqrt{1 - 0.25^2} = 0.968$$

$$\Rightarrow I_c = I_{o.c}\cos\varphi_{o.c} = (5)(0.25) = 1.25\,\text{A},\ I_m = I_{o.c}\sin\varphi_{o.c} = (5)(0.968) = 4.84\,\text{A}$$

1.16. From equation (1.53), we have:

$$R_{eq} = \frac{90}{10^2} = 0.9\,\Omega,\ |Z_{eq}| = \frac{25}{10} = 2.5\,\Omega,\ X_{eq} = \sqrt{2.5^2 - 0.9^2} = 2.33\,\Omega$$

1.17. From Section 1.10, we have:

Open-circuit test: $(\overbrace{10\,\text{kV}}^{V_{1n}}, 2\,\text{A}, \overbrace{500\,\text{W}}^{P_c})$

Short-circuit test: $(100\,\text{V}, \underbrace{10\,\text{A}}_{I_{1n}}, \underbrace{1000\,\text{W}}_{P_{cun}})$

One-third rated load: $P_c = \text{const.} = 500,\ P_{cu} = \frac{1}{3^2}(1000)$

$$\%\eta = \frac{\frac{1}{3}(100\text{k})(0.7)}{\frac{1}{3}(100\text{k})(0.7) + (500) + \frac{1}{9}(1000)}(100) = 97.45\%$$

1.18 From equation (1.45), we have:

$$\Rightarrow \begin{cases} (1): P_{c1} = k_h \dfrac{V_1^2}{f_1} + k_e V_1^2 \\ (2): P_{c2} = k_h \dfrac{V_2^2}{f_2} + k_e V_2^2 \end{cases} \Rightarrow (1)\times(V_2^2) + (2)\times(-V_1^2) \Rightarrow$$

$$k_h = \frac{P_{c1}V_2^2 - P_{c2}V_1^2}{\dfrac{V_1^2V_2^2}{f_1} - \dfrac{V_1^2V_2^2}{f_2}} = \frac{\dfrac{P_{c1}}{V_1^2} - \dfrac{P_{c2}}{V_2^2}}{\dfrac{1}{f_1} - \dfrac{1}{f_2}} = \frac{\left(\dfrac{P_{c1}}{V_1^2} f_1\right) f_2 - \left(\dfrac{P_{c2}}{V_2^2} f_2\right) f_1}{f_2 - f_1} = \frac{A_1 f_2 - A_2 f_1}{f_2 - f_1}$$

Where: $A_1 = \dfrac{P_{c1}}{V_1^2} f_1,\ A_2 = \dfrac{P_{c2}}{V_2^2} f_2$

$$(1) \Rightarrow k_e = \frac{P_{c1}}{V_1^2} - \frac{k_h}{f_1} = \frac{A_1}{f_1} - \frac{k_h}{f_1} = \frac{A_1}{f_1} - \frac{A_1 f_2 - A_2 f_1}{f_1(f_2 - f_1)} = \frac{A_1(f_2 - f_1) - A_1 f_2 + A_2 f_1}{f_1(f_2 - f_1)}$$

$$\Rightarrow k_e = \frac{A_2 - A_1}{f_2 - f_1}$$

1.19 From equation (1.53), we have:

$$\begin{cases} |Z_{eq1}| = \sqrt{R_{eq}^2 + X_{eq}^2} = \dfrac{V_{s.c}}{I_{s.c}} = \dfrac{100}{10} = 10\ \Omega \\ |Z_{eq2}| = \sqrt{R_{eq}^2 + (10X_{eq})^2} = \dfrac{V_{s.c}}{I_{s.c}} = \dfrac{100}{1.15} = 86.96\ \Omega \end{cases} \Rightarrow \begin{cases} R_{eq}^2 + X_{eq}^2 = 100\ \Omega \\ R_{eq}^2 + 100X_{eq}^2 = 7562\ \Omega \end{cases}$$

$$\Rightarrow \begin{cases} R = 4.962\ \Omega \\ X = 8.682\ \Omega \end{cases}$$

1.20 From equations (1.46) and (1.50), we have:

$$1)\ 0.98 = \frac{(1)(500\text{k})(0.8)}{(1)(500\text{k})(0.8) + P_c + P_{cun}} \Rightarrow P_c + P_{cun} = 8163$$

$$2)\ 0.99 = \frac{(0.5)(500\text{k})(1)}{(0.5)(500\text{k})(1) + P_c + \dfrac{P_{cun}}{4}} \Rightarrow P_c + \frac{P_{cun}}{4} = 2525$$

$$\Rightarrow P_c = 646\,\text{W}$$

2 Three-Phase Transformers

Part One: Lesson Summary

2.1 PRIMARY RELATIONSHIPS

In this section, the three-phase power equations are recalled [1].

Symmetric star–delta (Y–Δ) conversion:

$$Z_\Delta = 3Z_Y \tag{2.1}$$

Instantaneous power in three asymmetric phases:

$$P_{3\text{ph}}(t) = v_a(t)i_a(t) + v_b(t)i_b(t) + v_c(t)i_c(t) \tag{2.2}$$

A balanced three-phase voltage–time representation (equal magnitude and ±120° phase shift):

$$\begin{aligned} v_a(t) &= \sqrt{2}V\cos(\omega t + \theta_v) \\ v_b(t) &= \sqrt{2}V\cos(\omega t + \theta_v - 120) \\ v_c(t) &= \sqrt{2}V\cos(\omega t + \theta_v + 120) \end{aligned} \tag{2.3}$$

A balanced three-phase current–time representation (equal magnitude and ±120° phase shift):

$$\begin{aligned} i_a(t) &= \sqrt{2}I\cos(\omega t + \theta_i) \\ i_b(t) &= \sqrt{2}I\cos(\omega t + \theta_i - 120) \\ i_c(t) &= \sqrt{2}I\cos(\omega t + \theta_i + 120) \end{aligned} \tag{2.4}$$

A balanced three-phase voltage and current, phasor representation:

$$\begin{aligned} \hat{V}_a &= V\angle\theta_v, \hat{V}_b = V\angle(\theta_v - 120), \hat{V}_c = V\angle(\theta_v + 120) \\ \hat{I}_a &= I\angle\theta_i, \hat{I}_b = I\angle(\theta_i - 120), \hat{I}_c = I\angle(\theta_i + 120) \end{aligned} \tag{2.5}$$

Line-to-line voltage or simply line voltage:

$$\begin{aligned} \hat{V}_{ab} &= \hat{V}_a - \hat{V}_b = \sqrt{3}V\angle(\theta_v + 30) = V_L\angle(\theta_v + 30) \\ \hat{V}_{bc} &= \hat{V}_b - \hat{V}_c = \sqrt{3}V\angle(\theta_v - 90) = V_L\angle(\theta_v - 90) \\ \hat{V}_{ca} &= \hat{V}_c - \hat{V}_a = \sqrt{3}V\angle(\theta_v + 150) = V_L\angle(\theta_v + 150) \end{aligned} \tag{2.6}$$

DOI: 10.1201/9781003537564-2

For balanced three-phase voltages:

$$\hat{V}_a + \hat{V}_b + \hat{V}_c = 0, \quad \hat{V}_{ab} + \hat{V}_{bc} + \hat{V}_{ca} = 0 \tag{2.7}$$

The active power in a balanced three-phase system:

$$P_{3\text{ph}} = P_{\text{av}} = 3VI\cos(\varphi) = \sqrt{3}V_L I\cos(\varphi) = 3RI^2 \tag{2.8}$$

The reactive power in a balanced three-phase system:

$$Q_{3\text{ph}} = 3VI\sin(\varphi) = \sqrt{3}V_L I\sin(\varphi) = 3XI^2 \tag{2.9}$$

The complex power in a balanced three-phase system:

$$S_{3\text{ph}} = P_{3\text{ph}} + jQ_{3\text{ph}} = 3\hat{V}\hat{I}^* = 3ZI^2 = 3\frac{V^2}{Z^*} \tag{2.10}$$

The apparent power in a balanced three-phase system:

$$|S_{3\text{ph}}| = \sqrt{(P_{3\text{ph}})^2 + (Q_{3\text{ph}})^2} = 3VI = \sqrt{3}V_L I = 3|Z|I^2 = 3\frac{V^2}{|Z|} \tag{2.11}$$

2.2 IDEAL TRANSFORMER

Figures 2.1 and 2.2 depict two well-known three-phase transformer connections. Regardless of the transformer connection, the well-known equations (1.8) and (1.9) are determined as follows for each phase:

$$\frac{\hat{V}_A}{\hat{V}_a} = \frac{\hat{V}_B}{\hat{V}_b} = \frac{\hat{V}_C}{\hat{V}_c} = \frac{N_1}{N_2} \Rightarrow V \propto N \tag{2.12}$$

$$N_1\hat{I}_A + N_2\hat{I}_a = N_1\hat{I}_B + N_2\hat{I}_b = N_1\hat{I}_C + N_2\hat{I}_c = 0 \tag{2.13}$$

Figures 2.3 and 2.4 are typically used more than Figures 2.1 and 2.2 for the greater graphical presentation.

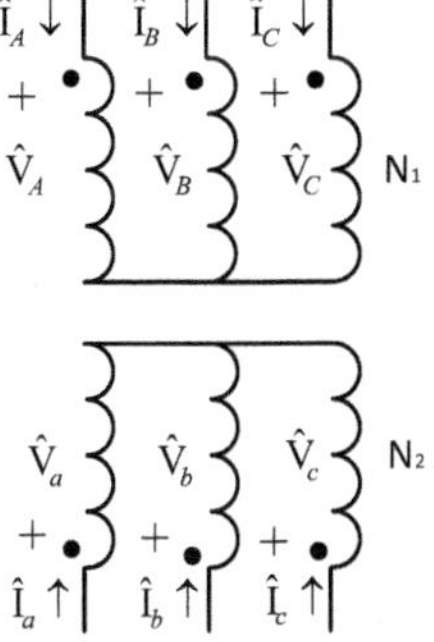

FIGURE 2.1 Ideal *Y–Y* transformer (the graphical show).

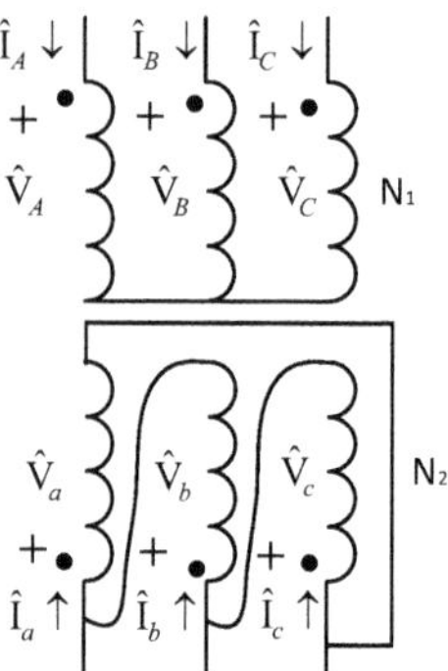

FIGURE 2.2 Ideal *Y*–Δ transformer (the graphical show).

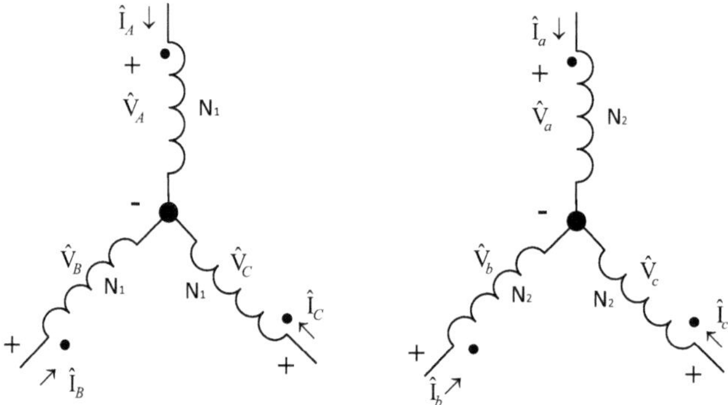

FIGURE 2.3 Ideal *Y*–*Y* transformer (the usual show).

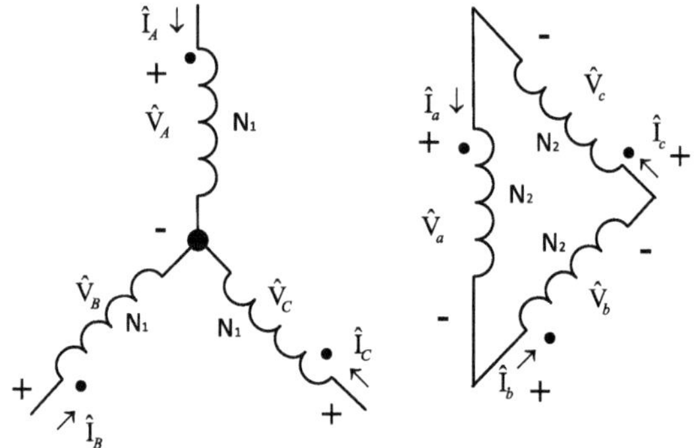

FIGURE 2.4 Ideal *Y*–Δ transformer (the usual show).

2.2.1 Types of Three-Phase Transformer Connections

A three-phase transformer typically combines delta and star connections. A combination of three-phase transformer connections is shown in Figure 2.5. It is important to note that the phases in the primary and secondary of the transformer are drawn in parallel for simplicity. It is for this reason that transformer connections are not shown in this book, as illustrated in Figure 2.6.

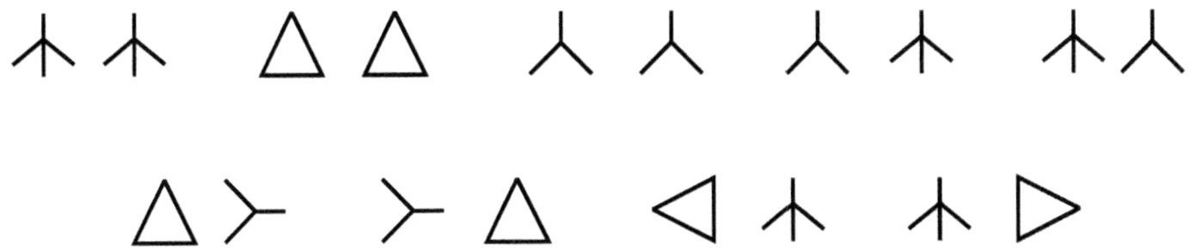

FIGURE 2.5 Different types of three-phase transformers (pay attention to the parallel sides).

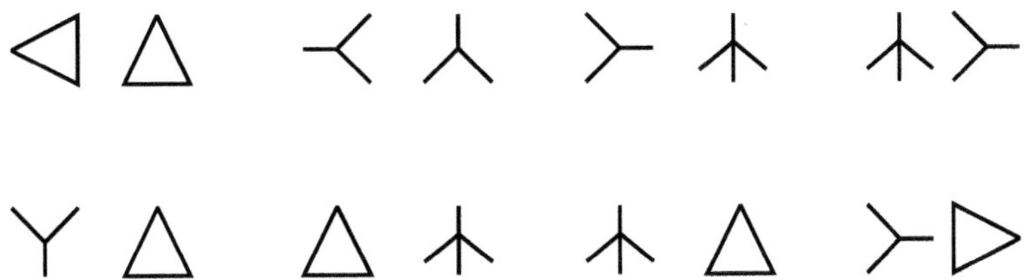

FIGURE 2.6 Different types of three-phase transformers (pay attention to the fact that the sides are not parallel).

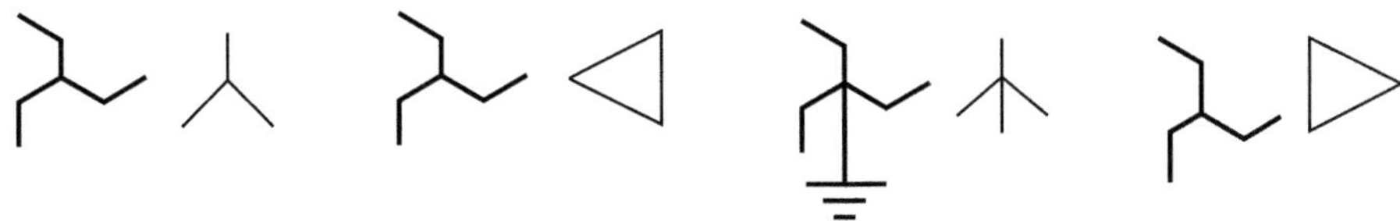

FIGURE 2.7 Different types of zigzag transformers (pay attention to the parallel sides)

2.2.2 Zigzag Three-Phase Transformer Connections

Zigzag connection transformers are relatively common in certain applications. However, their use is not universal, and other transformer configurations may be more appropriate for different scenarios. In fact, the zigzag connection combines a star and delta connection. A zigzag-connected transformer-winding configuration is being used for various reasons, such as providing a path for zero-sequence currents, eliminating harmonics, or reducing the effects of unbalanced loads. Figure 2.7 illustrates different types of zigzag transformers. A parallelism law for in-phase coils can also be seen in this figure.

Figure 2.8 displays the classic star-zigzag connection, and Figure 2.9 displays the typical graph. Here are the equations for zigzag connections:

$$\frac{\hat{V}_A}{\hat{V}_{a1}} = \frac{\hat{V}_A}{\hat{V}_{a2}} = \frac{\hat{V}_B}{\hat{V}_{b1}} = \frac{\hat{V}_B}{\hat{V}_{b2}} = \frac{\hat{V}_C}{\hat{V}_{c1}} = \frac{\hat{V}_C}{\hat{V}_{c2}} = \frac{N_1}{N_2} \Rightarrow V \propto N \tag{2.14}$$

$$N_1\hat{I}_A + N_2\hat{I}_{a1} + N_2\hat{I}_{a2} = 0 = N_1\hat{I}_B + N_2\hat{I}_{b1} + N_2\hat{I}_{b2} = N_1\hat{I}_C + N_2\hat{I}_{c1} + N_2\hat{I}_{c2} \tag{2.15}$$

2.2.3 Connections of Transformers

2.2.3.1 Series, Parallel, and Cascade

Calculations for symmetrical three-phase transformers are similar to for single-phase transformers. In the first chapter, Section 1.11, calculations and figures for these types of connections are given.

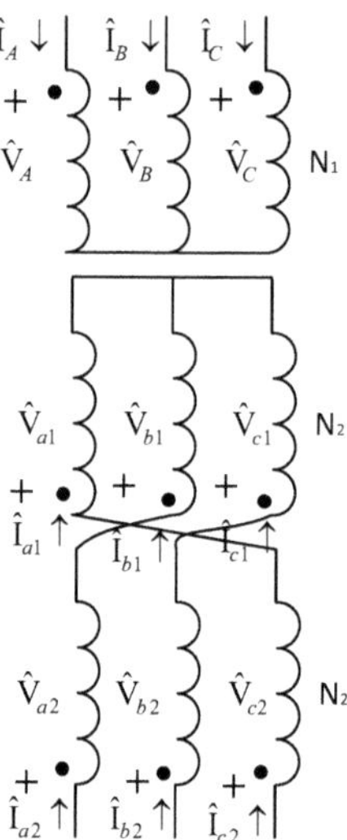

FIGURE 2.8 The classic star-zigzag connection.

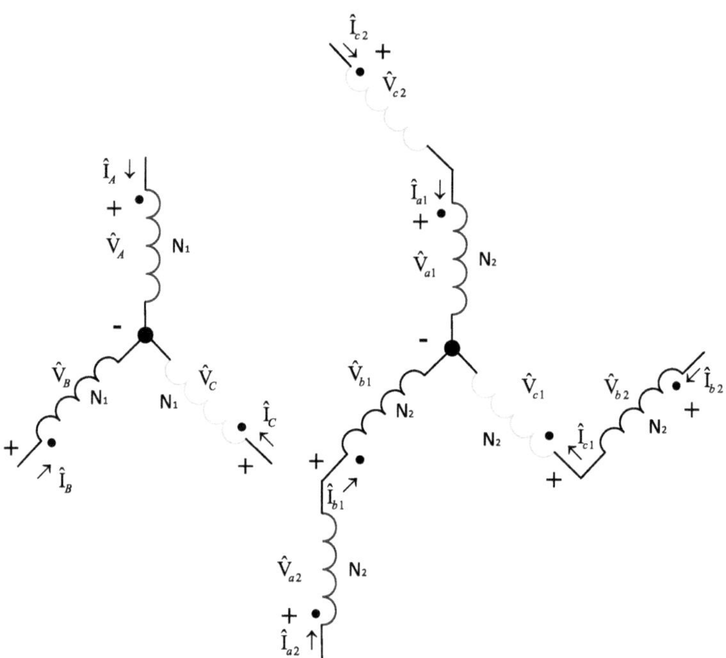

FIGURE 2.9 The typical graph of a star-zigzag connection.

2.2.3.1.1 Parallel Connection

In power system, instead of a large transformer, usually two smaller similar transformers are parallelized. Although, parallel transformers become more expensive, but they are being used because the outcome is more reliable in practice.

2.2.3.1.2 Series–Series

Due to the complexities involved in coordinating series transformers, ensuring proper voltage regulation, and increasing efficiency and minimizing losses, these configurations are not as widespread as other transformer arrangements.

2.2.3.1.3 Cascade Connection

For the ratings of the transformer over 500 kV, for economic reasons, a single transformer is not used. In these situations, two units are arranged in a cascade to produce the required voltage.

2.2.3.2 VV or Open-Delta Connection

Figure 2.10 shows a three-phase (V–V) or (open delta) transformer. The application of this type of connection is that a three-phase load can be supplied with only two coils (two single-phase transformers) without changing the voltage. In fact, power companies can initially use only two transformers for a large consumer's load and add a third transformer as the load increases.

Pay attention to the relationship between the voltages in the connection (delta–delta) and (open-delta) is constant and does not change.

$$\hat{V}_{AB} + \hat{V}_{BC} + \hat{V}_{CA} = 0, \quad \hat{V}_{ab} + \hat{V}_{bc} + \hat{V}_{ca} = 0 \tag{2.16}$$

The main difference in connection current or power (delta–delta) and (open-delta) is shown in Figure 2.11. In fact, the current and power (√3) times have decreased.

We also have:

$$\left|\frac{S_{\Delta\Delta}}{S_{o\Delta}}\right| = \frac{3VI}{\sqrt{3}VI} = \sqrt{3}, \quad \frac{P_{\Delta\Delta}}{P_{o\Delta}} = \frac{3VI\cos\varphi}{\sqrt{3}VI\cos\varphi} = \sqrt{3} \tag{2.17}$$

You can see the proof of Figure 2.11 and equation (2.17) in the solved problem 2.8.

2.2.3.3 Scott-T, (T–T) Connection

Scott-T transformer, or simply Scott connection, is a type of transformer used for transferring balanced two-phase electricity (with a 90° difference) from a balanced three-phase source (with a

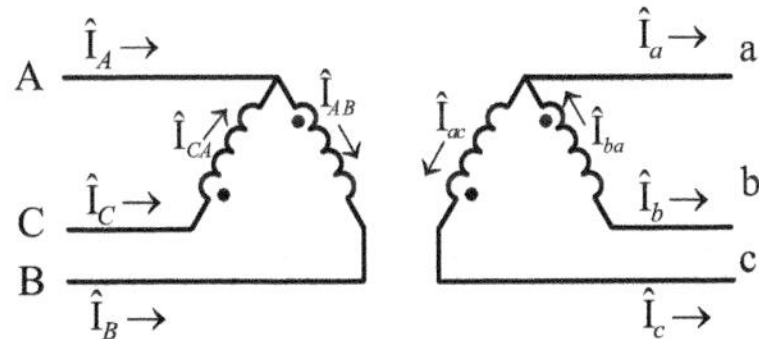

FIGURE 2.10 V–V or open-delta connection.

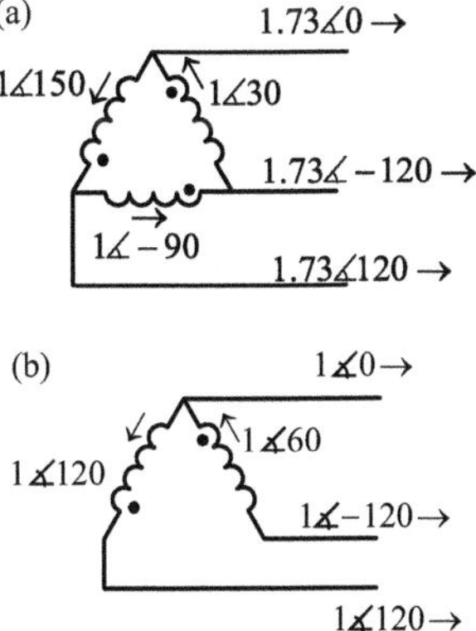

FIGURE 2.11 The main difference in connection between (a) delta–delta and (b) open-delta.

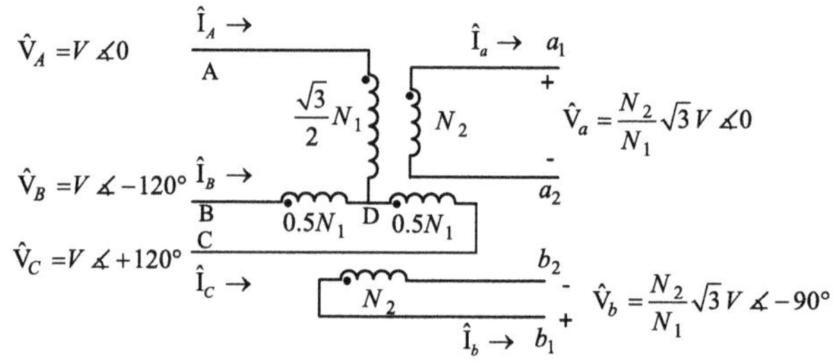

FIGURE 2.12 Scott-T, (T–T) connection.

120° angle difference), and vice versa. Westinghouse engineer Charles F. Scott developed the Scott transformer in the late 1890s. At that time, two-phase motor loads were also in use, and the Scott connection allowed them to be connected to newer three-phase supplies with equal currents in all three phases.

(See Figure 2.12) Note that the parallel connections are in phase. Balanced three-phase (*ABC*) becomes balanced two-phase(*ab*).

We also have:

$$S_{\text{out}} = \hat{V}_a \hat{I}_a^* + \hat{V}_b \hat{I}_b^* = 3VI\measuredangle\varphi = S_{\text{in}} \tag{2.18}$$

You can see the proof of Figure 2.12 and equation (2.18) in the solved problem 2.9. Also, you can prove the following relationship as an exercise if $a=(N_1/N_2)$.

$$\begin{bmatrix} \hat{V}_{AB} \\ \hat{V}_{BC} \\ \hat{V}_{CA} \end{bmatrix} = \begin{bmatrix} 0.5a\sqrt{3} & -0.5a \\ 0 & a \\ -0.5a\sqrt{3} & -0.5a \end{bmatrix} \begin{bmatrix} \hat{V}_a \\ \hat{V}_b \end{bmatrix}, \quad \begin{bmatrix} \hat{I}_A \\ \hat{I}_B \\ \hat{I}_C \end{bmatrix} = \frac{\sqrt{3}}{3} \begin{bmatrix} 2a^{-1} & 0 \\ -a^{-1} & a^{-1}\sqrt{3} \\ -a^{-1} & -a^{-1}\sqrt{3} \end{bmatrix} \begin{bmatrix} \hat{I}_a \\ \hat{I}_b \end{bmatrix} \tag{2.19}$$

2.3 UNBALANCED THREE-PHASE TRANSFORMER

To analyze the unbalanced transformer, the unbalanced voltages and currents must be converted into (+, –, 0) components. If $\alpha \triangleq 1\angle 120^\circ$, we have:

$$\overbrace{\begin{bmatrix} \hat{I}_a \\ \hat{I}_b \\ \hat{I}_c \end{bmatrix}}^{I^{abc}} = \overbrace{\begin{bmatrix} 1 & 1 & 1 \\ 1 & \alpha^2 & \alpha \\ 1 & \alpha & \alpha^2 \end{bmatrix}}^{A} \overbrace{\begin{bmatrix} \hat{I}_a^0 \\ \hat{I}_a^+ \\ \hat{I}_a^- \end{bmatrix}}^{I_a^{0+-}}, \quad \overbrace{\begin{bmatrix} \hat{I}_a^0 \\ \hat{I}_a^+ \\ \hat{I}_a^- \end{bmatrix}}^{I_a^{0+-}} = \frac{1}{3} \overbrace{\begin{bmatrix} 1 & 1 & 1 \\ 1 & \alpha & \alpha^2 \\ 1 & \alpha^2 & \alpha \end{bmatrix}}^{A^{-1}} \overbrace{\begin{bmatrix} \hat{I}_a \\ \hat{I}_b \\ \hat{I}_c \end{bmatrix}}^{I^{abc}} \tag{2.20}$$

Since the transformer is symmetrical, we have:

$$Z_T^+ = Z_T^- = Z_T^0 \tag{2.21}$$

Simply focus on the transformers' zero sequence model. The zero component of the models listed below is disconnected (Figure 2.13).

The zero component of the models listed below is connected (Figure 2.14).

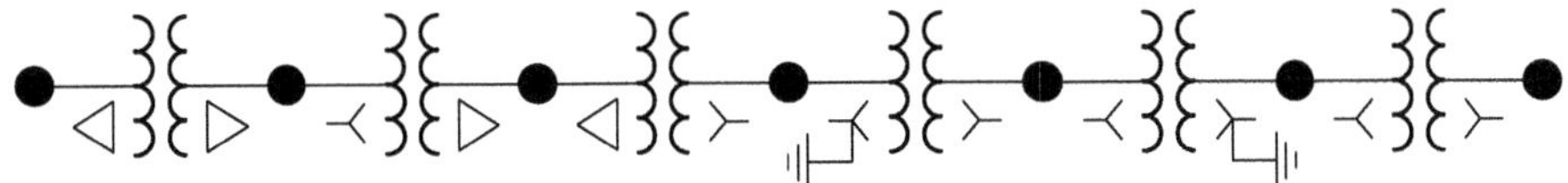

FIGURE 2.13 Disconnected models at zero components.

FIGURE 2.14 Connected models with zero components.

TABLE 2.1
Types of Transformer Testing

Transformer Vector Group	Short Circuit Impedance
Transformer ratio	Winding resistance
Load loss	Acoustic noise level
No-load loss and current	Harmonics of the no-load current
Insulation resistance	Power is taken by the fans and oil pumps
Dielectric tests	Buchholz relay
Temperature rise	Temperature indicators
On-load tap-changer	Pressure relief devices
Vacuum tests on tank	Oil preservation system

2.4 TRANSFORMER TESTS

Transformer tests can be divided into three categories: Type tests, routine tests, and special tests. See Table 2.1 and Chapter 5 for more information.

2.5 TRANSFORMER TRANSIENT MODELING

Sudden changes in voltage, current, or other electrical properties, known as transients, can pose significant challenges. Here's where transformer transient modeling comes to the forefront. Transient events can originate from various sources, including lightning strikes, switching operations (activating or deactivating transformers or other equipment), and short circuits. These events subject transformers to stresses beyond their typical steady-state operation. Unlike steady-state conditions where currents and voltages are relatively constant, transients can cause rapid fluctuations that can potentially damage the transformer's insulation, lead to mechanical stress, and ultimately reduce its lifespan.

Transformer transient modeling steps in to address these concerns. It involves creating mathematical representations of a transformer's behavior under transient conditions. These models consider the complex interactions between the transformer's electrical and magnetic properties. By simulating various transient scenarios through software, engineers can predict and analyze how a transformer will respond to these disturbances.

There are several approaches to transformer transient modeling, each with its own advantages. Equivalent circuit models offer a simplified approach by representing transformer components as basic electrical circuits. For more intricate analysis, the finite element method uses computer software to analyze the complex geometry of the transformer's core and windings. Specialized electromagnetic transient simulation programs provide a comprehensive approach, allowing engineers to simulate a wide range of transient events in detail.

The benefits of employing transformer transient modeling tools are numerous. By analyzing potential transient behavior, engineers can design transformers that are more robust and resilient to these disturbances. This translates into increased equipment lifespan and reduced maintenance costs. Additionally, these models can be used to optimize protection systems, such as surge arresters and circuit breakers, which safeguard transformers during faults. Ultimately, transformer transient modeling plays a crucial role in enhancing the overall reliability and stability of the power grid.

In conclusion, transformer transient modeling serves as an indispensable tool in the hands of power system engineers. By facilitating the prediction and analysis of transient behavior, it empowers them to design robust transformers, optimize protection schemes, and ensure the safe and reliable delivery of electricity to homes, businesses, and industries alike.

2.6 TRANSFORMER USES IN POWER SYSTEMS

Transformers are the backbone of the power grid. At power plants, transformers act as voltage boosters. Electricity generated is typically at a relatively low voltage, unsuitable for long-distance transmission due to energy loss in the form of heat. Step-up transformers significantly increase the voltage, minimizing these losses and enabling efficient power flow through transmission lines. As electricity nears consumption zones, step-down transformers take over. These transformers progressively reduce the voltage to levels safe and suitable for industries, businesses, and homes. This multi-stage voltage transformation ensures minimal energy loss during transmission while delivering electricity at appropriate voltages for end users [6–28].

Transformers not only manipulate voltage but also provide a degree of electrical isolation between circuits. This isolation helps protect sensitive equipment from voltage spikes or surges originating from other parts of the grid. Additionally, transformers can be used to convert single-phase to three-phase power or vice versa, catering to the diverse needs of different consumers.

2.7 UNDESIRABLE FEATURES OF THE THIRD HARMONIC

Third harmonic currents oscillate at three times the fundamental frequency, sabotaging transformer performance and efficiency. The primary cause of third harmonic distortion is the non-linear behavior of transformer cores. At higher operating levels, the core saturates, causing its permeability (resistance to magnetic fields) to vary. This non-linearity distorts the sinusoidal waveform of the incoming current, injecting unwanted third harmonic components [29–30].

The presence of third harmonics brings a symphony of undesirable consequences. Increased core losses translate into wasted energy and higher operating temperatures, shortening the transformer's lifespan. These harmonics also contribute to increased heating in the windings, further accelerating transformer degradation. Additionally, third harmonics can interact with the system impedance, causing voltage distortion and potential malfunctioning of sensitive electronic equipment.

Mitigating the impact of third harmonics requires a multi-pronged approach. Derating transformers, and operating them below their rated capacity, can minimize core saturation and reduce harmonic generation. Utilizing K-factor-rated transformers, designed to handle a specific level of harmonic distortion, can be another solution. K-factor-rated transformers are designed to handle non-linear loads more effectively, accommodating harmonics and reducing voltage distortion. In critical applications, harmonic filters can be employed to selectively remove the unwanted third harmonic frequencies.

In order to maintain system stability and transformer performance at their best, third harmonics must be avoided. By implementing appropriate mitigation strategies, engineers can maintain a harmonious flow of electricity, safeguarding transformers from premature aging and ensuring the smooth operation of the power grid.

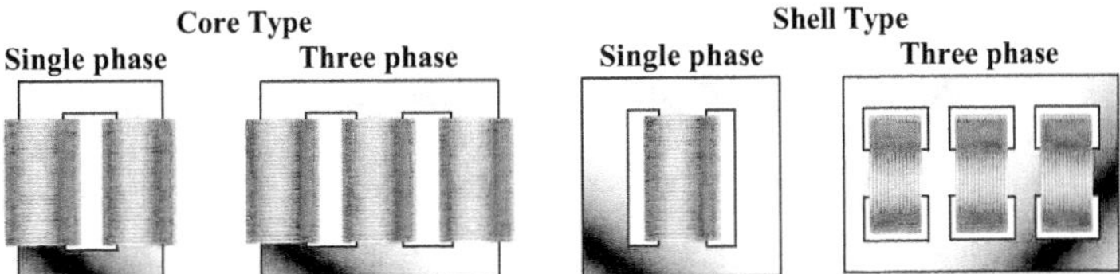

FIGURE 2.15 Core type and shell type transformers.

2.8 CORE-TYPE AND SHELL-TYPE TRANSFORMERS

Core-type and shell-type transformers refer to the arrangement of the core and windings within the transformer. In a core-type transformer, the windings surround a substantial part of the core, with the core in the center and the windings wound around it. In a shell-type transformer, the core surrounds the windings, with the windings being placed inside the core structure (See Figure 2.15).

The core type is more economical and easier to manufacture compared to the shell type. However, in power systems, the shell-type design is preferred for high-power rating and voltage applications. Because of its better short circuit strength characteristics. It also exhibits a better power-to-weight ratio.

2.9 THREE-LEG CORE, FIVE-LEG CORE

The two main core types are three-phase, three-leg, and five-leg. Here is a breakdown of their key differences.

2.9.1 Three-Legged Core

Frequently used in distribution class dry-type transformers – both low and medium voltages. It is the simplest and most common design. It consists of a central iron core leg (limb) with two outer limbs (yokes) on either side. Its most important advantage is that its production process is simple and leads to a reduction of production costs, and its most important disadvantage is higher core losses compared to the five-limb design. Because more magnetic flux has to travel a longer path in the core. Furthermore, it may experience high zero-sequence impedance, which can be a concern in unbalanced load conditions.

2.9.2 Five-Legged Core

Five-legged core designs are the standard for all distribution transformer applications today (regardless of whether or not is *Y–Y*). Since the cross-sectional area of the three inner limbs surrounded by the coils is double the size of the three-limb design, the cross-sectional area of the yoke and outer limbs can be half that of the inner limbs. This helps conserve material and reduce production costs as well. See Figure 2.16.

FIGURE 2.16 Three- and five-legged cores

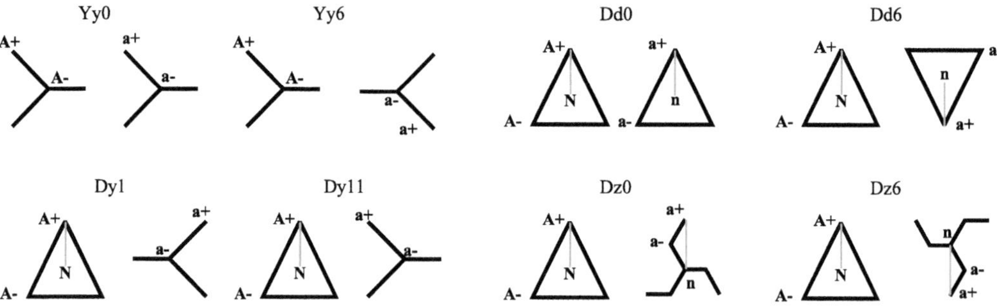

FIGURE 2.17 Vector groups.

The main advantages of the five-limb core are lower core losses due to a more efficient magnetic flux path, improved zero-sequence impedance, and useful in unbalanced load scenarios. Its main disadvantage is the more complicated manufacturing process, which leads to potentially higher costs, as well as less space available for winding per limb compared to a three-limb design.

Comparing the two types of cores, it can be said that three-limbed cores are favored for their simplicity and cost-effectiveness, while five-limb cores are preferred when lower core losses or improved zero-sequence impedance are crucial factors.

2.10 PHASE GROUPS

This is one of the important factors when paralleling several transformers together. The vector groups (n) of all possible three-phase transformer connections are categorized as follows:

Group 1 ($n = 0$): 0-degree phase displacement (Yy0, Dd0, Dz0).
Group 2 ($n = 6$): 180-degree phase displacement (Yy6, Dd6, Dz6).
Group 3 ($n = 1$): –30-degrees lag phase displacement (Dy1, Yd1, Yz1).
Group 4 ($n = 11$): +30-degrees lag phase displacement (Yd11, Dy11, Yz11).

See equation (2.22). ($\hat{V}_A$) is the primary voltage, ($\hat{V}_a$) is the secondary voltage, and (n) is the group vector.

$$\hat{V}_A = V_1 \measuredangle 0, \quad \hat{V}_a = V_2 \measuredangle -(n)(30^\circ) \tag{2.22}$$

Figure 2.17 shows the phasor diagram of some examples of famous vector groups.

2.11 THREE-WINDING TRANSFORMER STABILIZATION BY TERTIARY WINDING

Three-winding three-phase transformers, equipped with a special feature known as tertiary winding, play a crucial role in achieving network control. This additional winding acts as a conductor, harmonizing the electrical symphony within the transformer. The primary function of a three-phase transformer is to adjust voltage levels, either stepping up voltage for transmission lines or stepping down for distribution. However, challenges arise due to zero-sequence currents and third harmonic currents. Zero-sequence currents, caused by unbalanced loads or ground faults, lack a return path in some transformers, potentially leading to voltage instability. Third harmonic currents, generated by non-linear loads such as electronic devices, distort the sinusoidal waveform, affecting efficiency and equipment health.

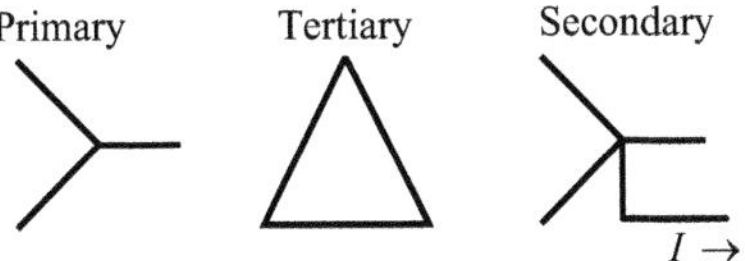

FIGURE 2.18 Three-phase transformer with tertiary winding.

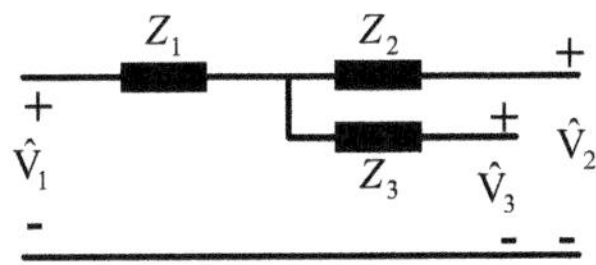

FIGURE 2.19 Single-phase simple equivalent circuit and perunit of three-winding transformer.

The tertiary winding, typically delta-connected, comes to the rescue. This additional winding provides a path for zero-sequence currents to circulate, preventing them from affecting the primary and secondary windings. Additionally, due to the inherent properties of delta connections, the third harmonic currents, in phase with each other, induce canceling magnetic fluxes in the tertiary winding. Consequently, these unwanted harmonics are not introduced to the primary or secondary sides, minimizing their disruptive effects.

The benefits of employing a tertiary winding extend beyond voltage stabilization. It can be used for reactive power compensation by connecting capacitors to the tertiary winding, which improves system voltage regulation. In some cases, the tertiary winding can also serve as a source of station service power, providing a dedicated power supply for equipment within the substation.

However, the inclusion of a tertiary winding adds complexity and cost to the transformer design. Therefore, the decision to utilize a tertiary winding depends on the specific application and the severity of zero-sequence current and third harmonic concerns.

(See Figure 2.18) Note that the zero component or the third harmonic can exist in the secondary and tertiary windings, but not in the primary winding. You can visualize the simple equivalent circuit and perunit of this transformer in Figure 2.19. You can easily calculate the impedances of this equivalent circuit by testing three short circuits in different coils.

Part Two: Answer-Question

2.12 TWO-CHOICE QUESTIONS (YES/NO)

1. When the load is balanced, the *YY* transformer operates satisfactorily.
2. The current ratio of the windings in a three-winding transformer is proportional to the reciprocal of the windings' turn ratio.
3. In three-phase transformers, grounding the *Y* connection is more common than using an ungrounded *Y* connection.
4. The zigzag-connected transformer-winding configuration is used for various applications to reduce the effects of unbalanced loads.
5. One larger transformer is usually used instead of two smaller, parallel, similar transformers in the power system.
6. Cascading transformers are a common practice for creating extremely high voltages needed for long-distance power transmission.
7. "Series–series" connection for transformers, where both primaries and secondaries are connected in series, is not a practical (commonly) used configuration.

8. Open delta transformer is for converting three phases to two phases.
9. Open delta transformer capacity is typically around 100% of a similarly sized three-phase transformer.
10. The widespread use of the Scott-T transformer has declined.
11. The Scott-T transformer remains a valuable tool for specific applications.
12. The three-phase input power of the ideal Scott-T transformer is equal to its two-phase output power.
13. In an ideal unbalanced three-phase transformer, the ratio of voltages across the windings is directly proportional to the ratio of the number of turns in the coils.
14. Unbalanced loads introduce unequal currents in the windings, causing deviations from the ideal inversely proportional relationship.
15. In an ideal unbalanced three-phase transformer, the ratio of currents in the windings is inversely proportional to the ratio of the number of turns in the coils.
16. One of the routine tests for transformers is the Buchholz relay test, which helps detect internal faults.
17. Sudden changes in voltage and current transformer, lead to mechanical stress and ultimately reduce its lifespan.
18. Transformers provide a degree of electrical isolation for decreasing loss.
19. Transformers ensure efficient power transmission over long distances.
20. Transformers deliver electricity at safe and usable voltages.
21. The primary culprit behind third harmonic distortion is the non-linear behavior of transformer cores.
22. The type of transformer connection has no effect on the third harmonic.
23. Three-leg core transformer has higher core losses compared to the five-limb design.
24. Five-limb cores are preferred when lower core losses or improved zero-sequence impedance are crucial factors.
25. Three-leg core transformer is more expensive compared to the five-limb design.
26. The vector groups are one of the important factors when parallel single-phase transformers together.
27. The phase displacement in a Yy6 transformer connection is 60 degrees.
28. In three-winding transformers, the zero component or the third harmonic can exist in the secondary and tertiary windings, but not in the primary winding.

2.13 KEY ANSWERS TO TWO-CHOICE QUESTIONS

Yes	1,3,4,6,7,10,11,12,13,14,16,17,19,20,21,23,24,28
No	2,5,8,9,15,18,22,25,26,27

2.14 DESCRIPTIVE QUESTIONS OF THE THREE-PHASE TRANSFORMERS

2.1 Determine all the voltages and currents in the symmetrical system shown in Figure 2.20, if we have:

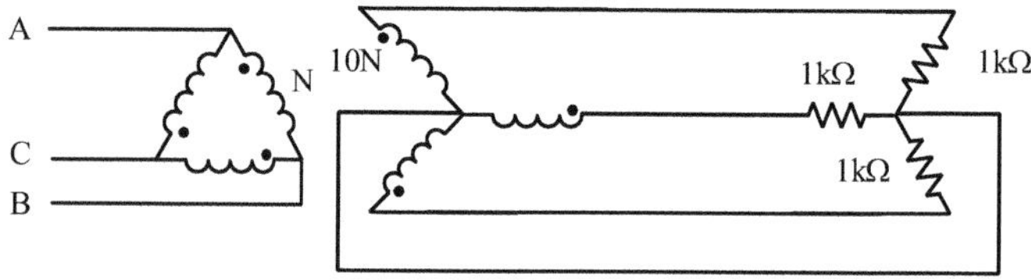

FIGURE 2.20 Question Figure 2.1.

$$\hat{V}_{AB} = 13.8\ \text{kV}\measuredangle 0,\quad \hat{V}_{BC} = 13.8\ \text{kV}\measuredangle -120°,\quad \hat{V}_{CA} = 13.8\ \text{kV}\measuredangle +120°$$

(Difficulty level ○ Easy ● Normal ○ Hard)

2.2 Determine all the voltages and currents in the symmetrical system shown in Figure 2.21, if we have:

$$\hat{V}_{AB} = 1000\sqrt{3}\ V\measuredangle 30°,\quad \hat{V}_{BC} = 1000\sqrt{3}\ V\measuredangle -90°,\quad \hat{V}_{CA} = 1000\sqrt{3}\ V\measuredangle +150°$$

$$Z = 100\measuredangle 30°$$

(Difficulty level ○ Easy ● Normal ○ Hard)

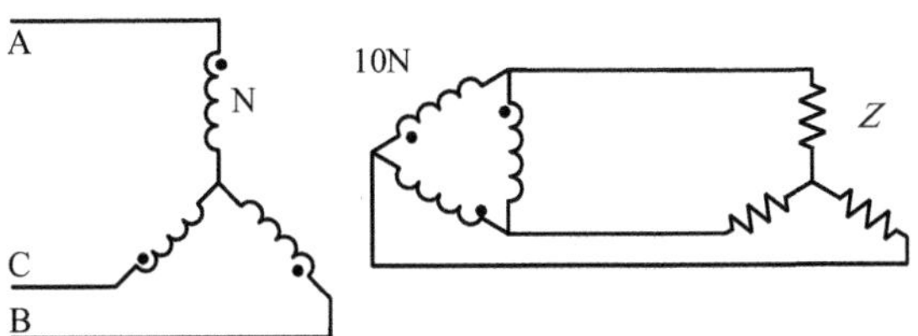

FIGURE 2.21 Question Figure 2.2.

2.3 Find $\frac{V_1}{V_2}$ in the symmetrical system shown in Figure 2.22.

(Difficulty level ● Easy ○ Normal ○ Hard)

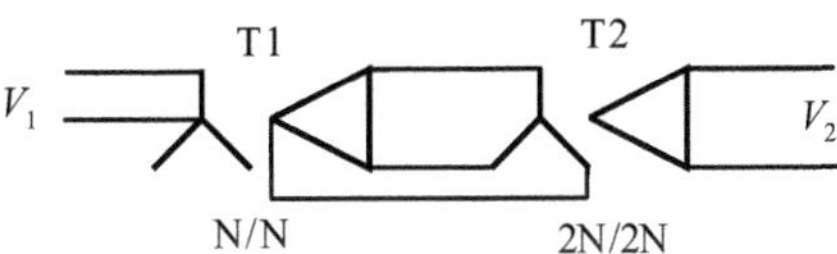

FIGURE 2.22 Question Figure 2.3.

2.4 Find $\frac{V_X}{V_{AB}}$ in the symmetrical system shown in Figure 2.23.

(Difficulty level ● Easy ○ Normal ○ Hard)

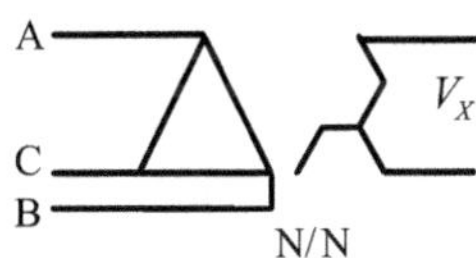

FIGURE 2.23 Question Figure 2.4.

2.5 Find the relationship between ($\hat{I}_1$) and ($\hat{I}_2$) in the symmetric system shown in Figure 2.24.

(Difficulty level ● Easy ○ Normal ○ Hard)

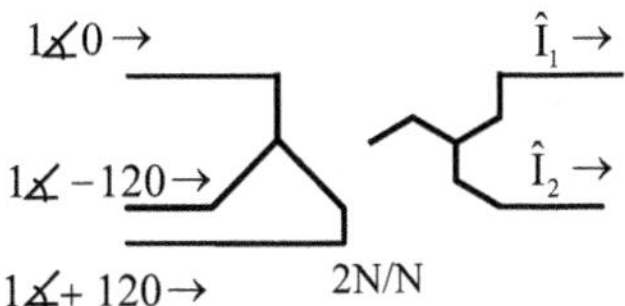

FIGURE 2.24 Question Figure 2.5.

2.6 Find $\hat{I}_1$ in the symmetric system shown in Figure 2.25.
(Difficulty level ● Easy ○ Normal ○ Hard)

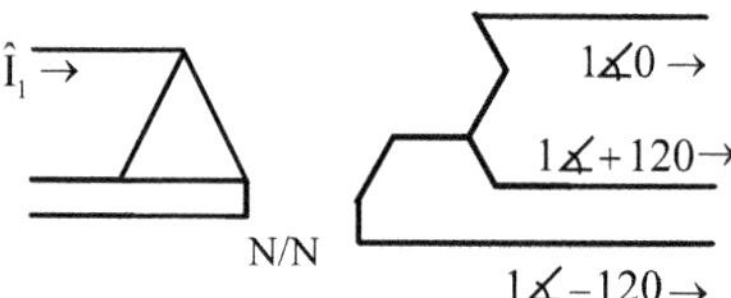

FIGURE 2.25 Question Figure 2.6.

2.7 Find $\frac{V_2}{V_1}$ in the symmetric system shown in Figure 2.26.
(Difficulty level ● Easy ○ Normal ○ Hard)

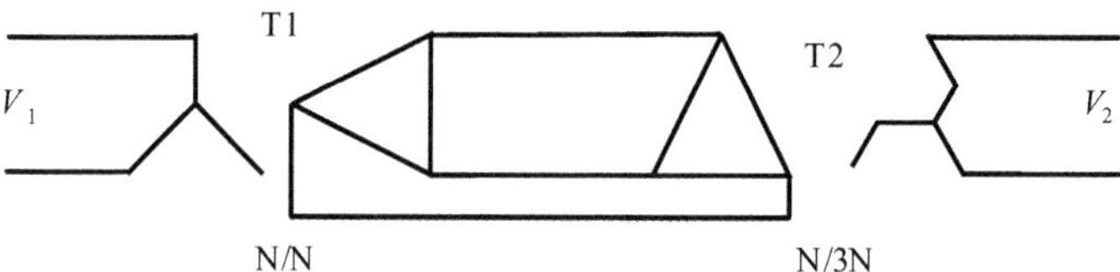

FIGURE 2.26 Question Figure 2.7.

2.8 Prove that the transmission power of the open-delta transformer, (√3) times is less than the normal delta-delta transformer.
(Difficulty level ○ Easy ○ Normal ● Hard)

2.9 By calculating the output current and voltage of the Scott transformer, prove that the output and input power are equal.
(Difficulty level ○ Easy ○ Normal ● Hard)

2.10 Consider 10 kVA, 20 kV/400 V transformer. (See Figure 2.27) Find the primary voltages and currents of the transformer. If:

$$\text{Load } A: (5\text{ kW}, 0.5\text{ lag}), \quad \text{Load } B: (8\text{ kW}, 0.5\text{ lag})$$

(Difficulty level ○ Easy ○ Normal ● Hard)

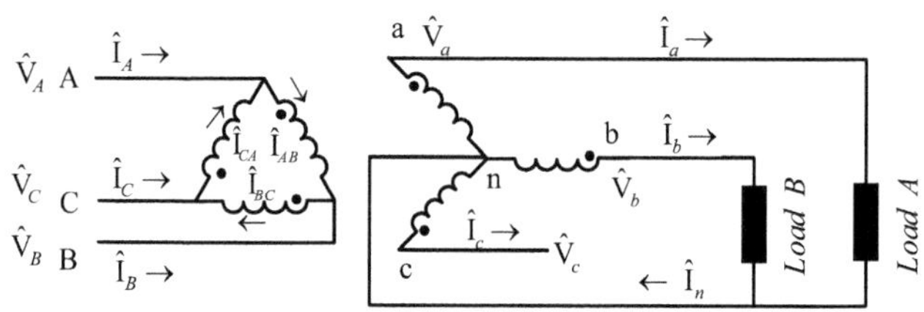

FIGURE 2.27 Question Figure 2.10.

2.11 In Figure 2.28, if the ratio of turns is one, find the primary current (I_A).
(Difficulty level ● Easy ○ Normal ○ Hard)

FIGURE 2.28 Question Figure 2.11.

2.15 DESCRIPTIVE ANSWERS OF THE THREE-PHASE TRANSFORMERS

2.1 See Figure 2.29.
We have:

$$\begin{cases} \hat{V}_{an} = \dfrac{10N}{10}\hat{V}_{AB} = 138\text{ kV}\measuredangle 0 \\ \hat{V}_{bn} = \dfrac{10N}{10}\hat{V}_{BC} = 138\text{ kV}\measuredangle -120^\circ \\ \hat{V}_{cn} = \dfrac{10N}{10}\hat{V}_{CA} = 138\text{ kV}\measuredangle +120^\circ \end{cases} \Rightarrow \begin{cases} \hat{I}_a = \dfrac{\hat{V}_{an}}{1\text{ k}\Omega} = 138\ A\measuredangle 0 \\ \hat{I}_b = \dfrac{\hat{V}_{bn}}{1\text{ k}\Omega} = 138\ A\measuredangle -120^\circ \\ \hat{I}_c = \dfrac{\hat{V}_{cn}}{1\text{ k}\Omega} = 138\ A\measuredangle +120^\circ \end{cases}$$

$$\Rightarrow \hat{I}_n = \hat{I}_a + \hat{I}_b + \hat{I}_c = 0$$

$$\Rightarrow \begin{cases} \hat{I}_{AB} = \dfrac{10N}{10}\hat{I}_a = 1.38\text{ kA}\measuredangle 0 \\ \hat{I}_{BC} = \dfrac{10N}{10}\hat{I}_b = 1.38\text{ kA}\measuredangle -120^\circ \\ \hat{I}_{CA} = \dfrac{10N}{10}\hat{I}_c = 1.38\text{ kA}\measuredangle +120^\circ \end{cases} \Rightarrow \begin{cases} \hat{I}_A = \hat{I}_{AB} - \hat{I}_{CA} = 1.38\sqrt{3}\text{ kA}\measuredangle -30^\circ \\ \hat{I}_B = \hat{I}_{BC} - \hat{I}_{AB} = 1.38\sqrt{3}\text{ kA}\measuredangle -150^\circ \\ \hat{I}_C = \hat{I}_{CA} - \hat{I}_{BC} = 1.38\sqrt{3}\text{ kA}\measuredangle +90^\circ \end{cases}$$

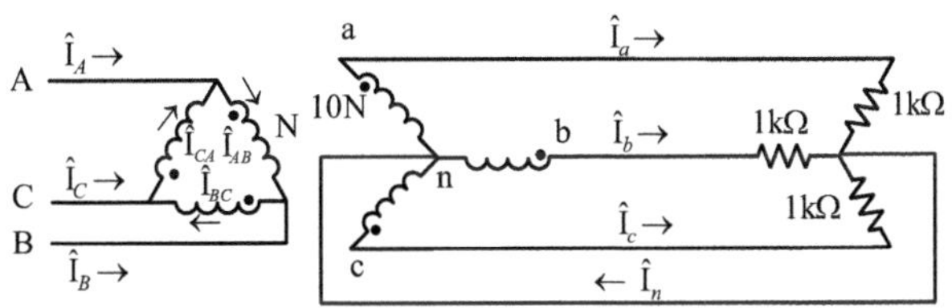

FIGURE 2.29 Answer Figure 2.1

2.2 See Figure 2.30.
We have:

$$\begin{cases} \hat{V}_{AB} = 1000\sqrt{3}\text{ V}\measuredangle 30^\circ \\ \hat{V}_{BC} = 1000\sqrt{3}\text{ V}\measuredangle -90^\circ \\ \hat{V}_{CA} = 1000\sqrt{3}\text{ V}\measuredangle +150^\circ \end{cases} \underset{\text{Angle}(-30^\circ)}{\overset{\text{Magnitude}(\div\sqrt{3})}{\Longrightarrow}} \begin{cases} \hat{V}_A = 1\text{ kV}\measuredangle 0^\circ \\ \hat{V}_B = 1\text{ kV}\measuredangle -120^\circ \\ \hat{V}_C = 1\text{ kV}\measuredangle +120^\circ \end{cases} \overset{\text{Transformer}}{\Longrightarrow}$$

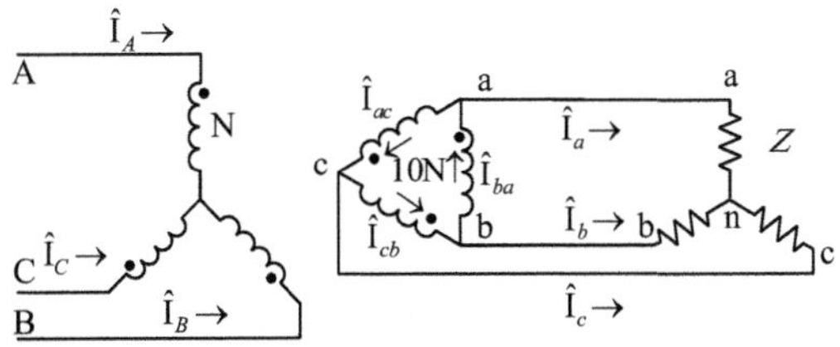

FIGURE 2.30 Answer Figure 2.2.

$$\begin{cases} \hat{V}_{ab} = 10\hat{V}_A = 10\ \text{kV}\measuredangle 0^\circ \\ \hat{V}_{bc} = 10\hat{V}_B = 10\ \text{kV}\measuredangle -120^\circ \\ \hat{V}_{ca} = 10\hat{V}_C = 10\ \text{kV}\measuredangle +120^\circ \end{cases} \xrightarrow[\text{Angle}(-30^\circ)]{\text{Magnitude}(\div\sqrt{3})} \begin{cases} \hat{V}_a = 10\dfrac{\sqrt{3}}{3}\ \text{kV}\measuredangle -30^\circ \\ \hat{V}_b = 10\dfrac{\sqrt{3}}{3}\ \text{kV}\measuredangle -150^\circ \\ \hat{V}_c = 10\dfrac{\sqrt{3}}{3}\ \text{kV}\measuredangle +90^\circ \end{cases} \overset{\text{Load}}{\Rightarrow}$$

$$\begin{cases} \hat{I}_a = \dfrac{\hat{V}_a}{Z} = 100\dfrac{\sqrt{3}}{3}\ A\measuredangle -60^\circ \\ \hat{I}_b = \dfrac{\hat{V}_b}{Z} = 100\dfrac{\sqrt{3}}{3}\ A\measuredangle -180^\circ \\ \hat{I}_c = \dfrac{\hat{V}_c}{Z} = 100\dfrac{\sqrt{3}}{3}\ A\measuredangle +60^\circ \end{cases} \xrightarrow[\text{Angle}(+30^\circ)]{\text{Magnitude}(\div\sqrt{3})\ \text{Why?}} \begin{cases} \hat{I}_{ba} = \dfrac{100}{3}\ A\measuredangle -30^\circ \\ \hat{I}_{cb} = \dfrac{100}{3}\ A\measuredangle -150^\circ \\ \hat{I}_{ac} = \dfrac{100}{3}\ A\measuredangle +90^\circ \end{cases} \overset{\text{Transformer}}{\Rightarrow}$$

$$\Rightarrow \begin{cases} \hat{I}_A = 10\hat{I}_{ba} = \dfrac{1000}{3}\ A\measuredangle -30^\circ \\ \hat{I}_B = 10\hat{I}_{cb} = \dfrac{1000}{3}\ A\measuredangle -150^\circ \\ \hat{I}_C = 10\hat{I}_{ac} = \dfrac{1000}{3}\ A\measuredangle +90^\circ \end{cases}$$

Now we compare the input and output powers, which must be equal.

$$S_{\text{in}} = \hat{V}_A(\hat{I}_A)^* + \hat{V}_B(\hat{I}_B)^* + \hat{V}_C(\hat{I}_C)^* = 3\hat{V}_A(\hat{I}_A)^* = 1\ \text{MVA}\measuredangle 30^\circ$$

$$S_{\text{out}} = \hat{V}_a(\hat{I}_a)^* + \hat{V}_b(\hat{I}_b)^* + \hat{V}_c(\hat{I}_c)^* = 3\hat{V}_a(\hat{I}_a)^* = 1\ \text{MVA}\measuredangle 30^\circ$$

2.3 See Figure 2.31.

From T2, we have: $\dfrac{V_4}{V_2} = \dfrac{2N}{2N} = 1$

From T1, we have: $\dfrac{V_1}{V_3} = \dfrac{N}{N} = 1$

Connection between (T1) and (T2), we have: $V_3 = V_4\sqrt{3}$

$$\Rightarrow \frac{V_1}{V_2} = \frac{V_1}{V_3}\frac{V_3}{V_4}\frac{V_4}{V_2} = (1)(\sqrt{3})(1) = \sqrt{3}$$

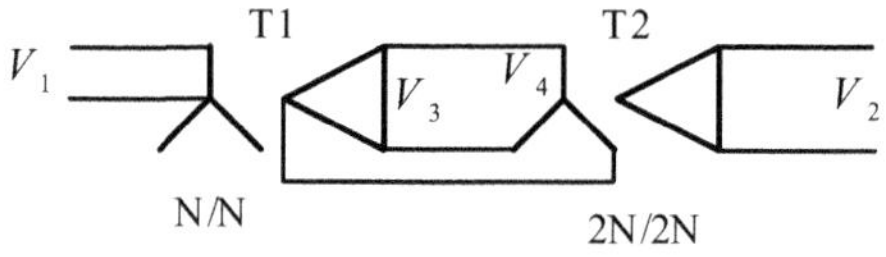

FIGURE 2.31 Answer Figure 2.3.

2.4 See Figure 2.32.

$$\text{KVL}: V_X = V_{AB} - V_{CA} + V_{AB} - V_{BC} = 2V_{AB} - (V_{CA} + V_{BC}) = 3V_{AB}$$

$$\Rightarrow \frac{V_L^{\text{Zigzag}}}{V_L^{\Delta}} = 3$$

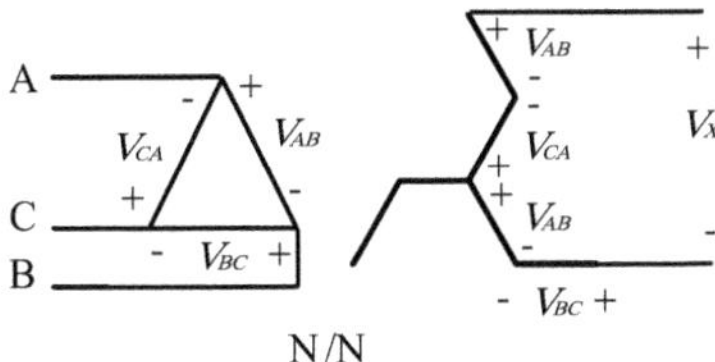

FIGURE 2.32 Answer Figure 2.4.

2.5 (See Figure 2.33) Parallel coils are in phase and we have:
Note that the current must enter the dot point.

$$2N(1\measuredangle 0)+N(-\hat{I}_1)+N(+\hat{I}_2)=0\Rightarrow \hat{I}_1-\hat{I}_2=2\measuredangle 0$$

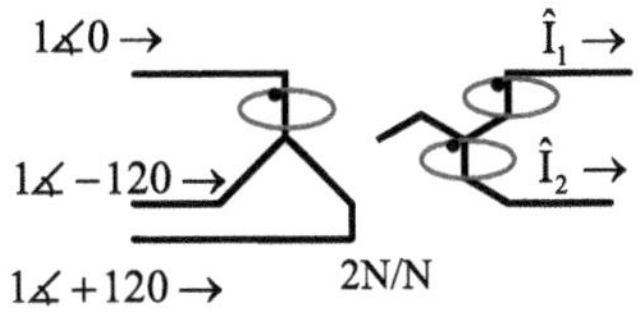

$1\measuredangle 0\rightarrow$ $\hat{I}_1\rightarrow$
$1\measuredangle -120\rightarrow$ $\hat{I}_2\rightarrow$
$1\measuredangle +120\rightarrow$ 2N/N

FIGURE 2.33 Answer Figure 2.5.

2.6 (See Figure 2.34) Parallel coils are in phase and we have:
Note that the current must enter the dot point. From equation (1.9), we have:

$$N(\hat{I}_A)+N(-1\measuredangle 0)+N(1\measuredangle 120)=0\Rightarrow \hat{I}_A=\sqrt{3}\measuredangle -30^\circ$$

$$N(\hat{I}_B)+N(1\measuredangle 0)+N(-1\measuredangle -120)=0\Rightarrow \hat{I}_B=\sqrt{3}\measuredangle -150^\circ$$

$$\Rightarrow \hat{I}_1=\hat{I}_A-\hat{I}_B=3\measuredangle 0^\circ$$

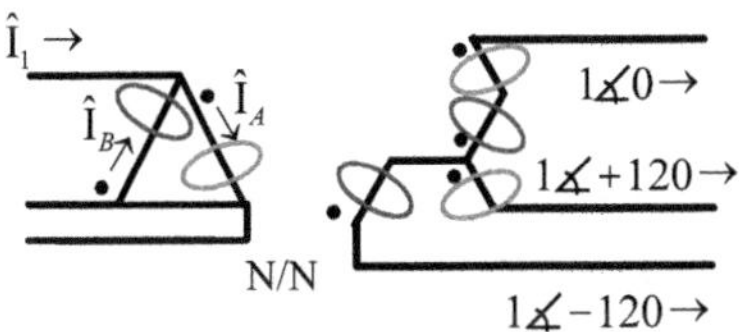

FIGURE 2.34 Answer Figure 2.6.

2.7 (See Figure 2.35) Using the ratio of the coil turns, we compute the voltages from the left to the right.
1. Line voltage to phase voltage
2. Voltage transfer with transformer conversion ratio
3. KVL : $V_{ab}=V_{bc}=V_{ca}=\dfrac{V_1}{\sqrt{3}}$

$$V_2=(3)V_{ab}-(3)V_{ca}+(3)V_{ab}-(3)V_{bc}=(6)V_{ab}-(3)(V_{ca}+V_{bc})=(9)V_{ab}$$

4.
$$\Rightarrow \frac{V_2}{V_1}=\frac{V_2}{V_{ab}}\frac{V_{ab}}{V_1}=(9)\left(\frac{1}{\sqrt{3}}\right)=3\sqrt{3}$$

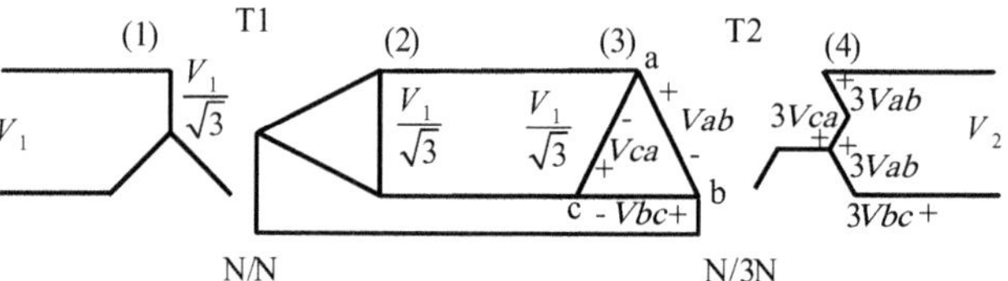

FIGURE 2.35 Answer Figure 2.7.

2.8 See Figure 2.36.

First, we write the (Δ–Δ) transformer output equations. Suppose (Figure 2.36a):

$$\begin{cases} \hat{V}_a = V\measuredangle 0 \\ \hat{V}_b = V\measuredangle -120° \\ \hat{V}_c = V\measuredangle +120° \end{cases}, \quad \begin{cases} \hat{I}_a = I\measuredangle -\varphi \\ \hat{I}_b = I\measuredangle -\varphi - 120° \\ \hat{I}_c = I\measuredangle -\varphi + 120° \end{cases}$$

Then:

$$\Rightarrow \begin{cases} \hat{V}_{ab} = \sqrt{3}V\measuredangle 30° \\ \hat{V}_{bc} = \sqrt{3}V\measuredangle -90° \\ \hat{V}_{ca} = \sqrt{3}V\measuredangle +150° \end{cases}, \quad \begin{cases} \hat{I}_{ba} = \frac{1}{\sqrt{3}} I\measuredangle -\varphi + 30° \\ \hat{I}_{cb} = \frac{1}{\sqrt{3}} I\measuredangle -\varphi - 90° \\ \hat{I}_{ac} = \frac{1}{\sqrt{3}} I\measuredangle -\varphi + 150° \end{cases}$$

Then:

$$S_{\text{out}}^{\Delta\Delta} = \hat{V}_a(\hat{I}_a)^* + \hat{V}_b(\hat{I}_b)^* + \hat{V}_c(\hat{I}_c)^* = \hat{V}_{ab}(\hat{I}_{ba})^* + \hat{V}_{bc}(\hat{I}_{cb})^* + \hat{V}_{ca}(\hat{I}_{ac})^* = 3VI\measuredangle\varphi$$

$$P_{\text{out}}^{\Delta\Delta} = 3VI\cos\varphi, \quad Q_{\text{out}}^{\Delta\Delta} = 3VI\sin\varphi$$

Second, we write the (Open-Δ) transformer output equations. Because the voltage has a direct relationship with the number of turns, there is only one unknown voltage that is calculated from KVL. We have (Figure 2.36b):

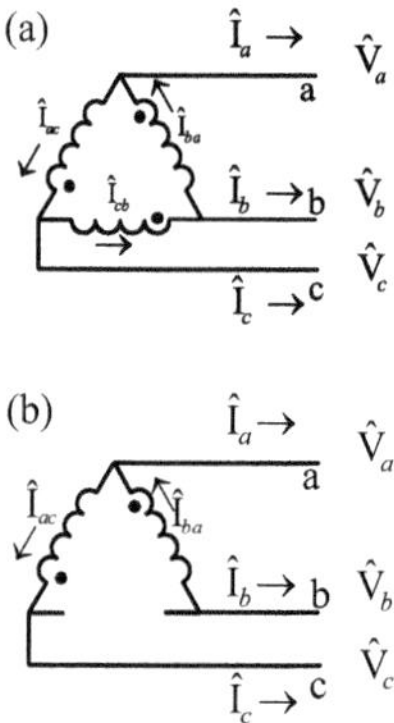

FIGURE 2.36 Answer Figure 2.8. (a) Normal delta connection and (b) Open-delta connection.

$$\begin{cases} \hat{V}_{ab} = \sqrt{3}V\measuredangle 30^\circ \\ \hat{V}_{bc} = ? \\ \hat{V}_{ca} = \sqrt{3}V\measuredangle +150^\circ \end{cases}, \quad \hat{V}_{ab} + \hat{V}_{bc} + \hat{V}_{ca} = 0 \Rightarrow \hat{V}_{bc} = -\left(\hat{V}_{ab} + \hat{V}_{ca}\right) = \sqrt{3}V\measuredangle -90^\circ$$

That is, the voltages in (open-Δ) remain unchanged.

$$\Rightarrow \begin{cases} \hat{V}_{ab} = \sqrt{3}\,V\measuredangle 30^\circ \\ \hat{V}_{bc} = \sqrt{3}\,V\measuredangle -90^\circ \\ \hat{V}_{ca} = \sqrt{3}\,V\measuredangle +150^\circ \end{cases} \Rightarrow \begin{cases} \hat{V}_a = V\measuredangle 0 \\ \hat{V}_b = V\measuredangle -120^\circ \\ \hat{V}_c = V\measuredangle +120^\circ \end{cases}$$

Also, to compare the currents, the nominal value of the current of the coils has not changed, so we have:

$$\begin{cases} \hat{I}_{ba} = \frac{1}{\sqrt{3}} I\measuredangle\theta_{ba} \\ \hat{I}_{cb} = 0 \\ \hat{I}_{ac} = \frac{1}{\sqrt{3}} I\measuredangle\theta_{ac} \end{cases} \tag{2.23}$$

We also know that with balanced voltages and symmetrical loads, the currents are also balanced, so we have:

$$\begin{cases} \hat{I}_a = I_a\measuredangle -\varphi \\ \hat{I}_b = I_b\measuredangle -\varphi - 120^\circ \\ \hat{I}_c = I_c\measuredangle -\varphi + 120^\circ \end{cases} \tag{2.24}$$

With KCL in Figure 2.36b, we have:

$$\begin{cases} 1)\ \hat{I}_{ba} = -\hat{I}_b \\ 2)\ \hat{I}_{ac} = \hat{I}_c \\ 3)\ \hat{I}_a = \hat{I}_{ba} - \hat{I}_{ac} \end{cases} \tag{2.25}$$

Then from equations (2.23–2.25), we have:

1. $\hat{I}_{ba} = -\hat{I}_b \Rightarrow \frac{I}{\sqrt{3}}\measuredangle\theta_{ba} = -I_b\measuredangle(-\varphi - 120^\circ) = I_b\measuredangle(-\varphi - 120^\circ + 180^\circ) = I_b\measuredangle(-\varphi + 60^\circ)$

   $$\Rightarrow \hat{I}_{ba} = \frac{I}{\sqrt{3}}\measuredangle(-\varphi + 60^\circ),\ \hat{I}_b = \frac{I}{\sqrt{3}}\measuredangle(-\varphi - 120^\circ)$$

2. $\hat{I}_{ac} = \hat{I}_c \Rightarrow \frac{1}{\sqrt{3}} I\measuredangle\theta_{ac} = I_c\measuredangle -\varphi + 120^\circ$

   $$\Rightarrow \hat{I}_{ac} = \frac{1}{\sqrt{3}} I\measuredangle(-\varphi + 120^\circ),\quad \hat{I}_c = \frac{1}{\sqrt{3}} I\measuredangle(-\varphi + 120^\circ)$$

3. $\hat{I}_a = \hat{I}_{ba} - \hat{I}_{ac} \Rightarrow \hat{I}_a = \frac{I}{\sqrt{3}}\measuredangle(-\varphi+60°) - \frac{I}{\sqrt{3}}\measuredangle(-\varphi+120°) = \frac{I}{\sqrt{3}}\measuredangle-\varphi \Rightarrow$

In short, the voltages and currents in Open-Δ are as follows:

$$\begin{cases} \hat{V}_{ab} = \sqrt{3}V\measuredangle 30° \\ \hat{V}_{bc} = \sqrt{3}V\measuredangle -90° \\ \hat{V}_{ca} = \sqrt{3}V\measuredangle +150° \end{cases}, \begin{cases} \hat{I}_{ba} = \frac{I}{\sqrt{3}}\measuredangle -\varphi+60° \\ \hat{I}_{cb} = 0 \\ \hat{I}_{ac} = \frac{I}{\sqrt{3}}\measuredangle -\varphi+120° \end{cases}, \begin{cases} \hat{V}_a = V\measuredangle 0 \\ \hat{V}_b = V\measuredangle -120° \\ \hat{V}_c = V\measuredangle +120° \end{cases}, \begin{cases} \hat{I}_a = \frac{I}{\sqrt{3}}\measuredangle -\varphi \\ \hat{I}_b = \frac{I}{\sqrt{3}}\measuredangle -\varphi-120° \\ \hat{I}_c = \frac{I}{\sqrt{3}}\measuredangle -\varphi+120° \end{cases}$$

Then:

$$S_{\text{out}}^{o\Delta} = \hat{V}_a(\hat{I}_a)^* + \hat{V}_b(\hat{I}_b)^* + \hat{V}_c(\hat{I}_c)^* = \hat{V}_{ab}(\hat{I}_{ba})^* + \hat{V}_{bc}(\hat{I}_{cb})^* + \hat{V}_{ca}(\hat{I}_{ac})^* = 3V\frac{I}{\sqrt{3}}\measuredangle\varphi$$

$$\Rightarrow S_{\text{out}}^{o\Delta} = \sqrt{3}VI\measuredangle\varphi, \quad P_{\text{out}}^{o\Delta} = \sqrt{3}VI\cos\varphi, \quad Q_{\text{out}}^{o\Delta} = \sqrt{3}VI\sin\varphi$$

$$\Rightarrow \left|\frac{S_{\Delta\Delta}}{S_{o\Delta}}\right| = \frac{3VI}{\sqrt{3}VI} = \sqrt{3}, \quad \frac{P_{\Delta\Delta}}{P_{o\Delta}} = \frac{3VI\cos\varphi}{\sqrt{3}VI\cos\varphi} = \sqrt{3}$$

2.9 See Figure 2.12.

Voltages:

Three balanced phases are assumed as follows:

$$\hat{V}_A = V\measuredangle 0, \quad \hat{V}_B = V\measuredangle -120°, \quad \hat{V}_C = V\measuredangle +120° \tag{2.26}$$

From the relation of the transformer, we have:

$$1)\ \frac{\hat{V}_a}{\hat{V}_{AD}} = \frac{N_2}{0.5\sqrt{3}N_1}, \quad 2)\ \frac{\hat{V}_b}{\hat{V}_{BD}} = \frac{\hat{V}_b}{\hat{V}_{DC}} = \frac{N_2}{0.5N_1}, \quad 3)\ \hat{V}_{BD} = \hat{V}_{DC} \tag{2.27}$$

From KVL:

$$\begin{gathered} \hat{V}_B - \hat{V}_C = \hat{V}_{BD} + \hat{V}_{DC} = 2\hat{V}_{BD} \Rightarrow \\ \hat{V}_{BD} = 0.5(V\measuredangle -120° - V\measuredangle 120°) = 0.5\sqrt{3}\,V\measuredangle -90° \end{gathered} \tag{2.28}$$

From equations (2.27) and (2.28), we have:

$$\hat{V}_b = \frac{N_2}{0.5N_1}0.5\sqrt{3}\,V\measuredangle -90° \Rightarrow \hat{V}_b = \frac{N_2}{N_1}\sqrt{3}\,V\measuredangle -90° \tag{2.29}$$

From KVL and equation (2.28):

$$\begin{gathered} \hat{V}_A - \hat{V}_B = \hat{V}_{AD} + \hat{V}_{DB} \Rightarrow \hat{V}_{AD} = (V\measuredangle 0) - (V\measuredangle -120°) - 0.5\sqrt{3}\,V\measuredangle +90° \\ \Rightarrow \hat{V}_{AD} = 1.5\,V\measuredangle 0 \end{gathered} \tag{2.30}$$

Then from equations (2.27) and (2.30), we have:

$$\hat{V}_a = \frac{N_2}{0.5\sqrt{3}N_1}1.5\,V\measuredangle 0 \Rightarrow \hat{V}_a = \frac{N_2}{N_1}\sqrt{3}\,V\measuredangle 0 \tag{2.31}$$

Currents:

Three balanced phases are assumed as follows:

$$\hat{I}_A = I\measuredangle -\varphi, \quad \hat{I}_B = I\measuredangle -\varphi - 120°, \quad \hat{I}_C = I\measuredangle -\varphi + 120° \tag{2.32}$$

From the relation of the transformer, we have:

$$1)\ 0.5\sqrt{3}N_1(\hat{I}_A) + N_2(-\hat{I}_a) = 0 \Rightarrow \hat{I}_a = 0.5\sqrt{3}\frac{N_1}{N_2}\hat{I}_A = 0.5\sqrt{3}\frac{N_1}{N_2}(I\measuredangle -\varphi) \tag{2.33}$$

$$2)\ 0.5N_1(\hat{I}_B) + 0.5N_1(-\hat{I}_C) + N_2(-\hat{I}_b) = 0 \Rightarrow \hat{I}_b = 0.5\frac{N_1}{N_2}(\hat{I}_B - \hat{I}_C) \tag{2.34}$$

From equations (2.32) and (2.34), we have:

$$\hat{I}_b = 0.5\frac{N_1}{N_2}\sqrt{3}I\measuredangle -\varphi - 90 \tag{2.35}$$

Powers:

The output power of the Scott transformer can be calculated as follows:

$$S_{\text{out}} = \hat{V}_a\hat{I}_a^* + \hat{V}_b\hat{I}_b^* = \left(\frac{N_2}{N_1}\sqrt{3}\,V\measuredangle 0\right)\left(0.5\sqrt{3}\frac{N_1}{N_2}I\measuredangle -\varphi\right)^*$$

$$+\left(\frac{N_2}{N_1}\sqrt{3}\,V\measuredangle -90°\right)\left(0.5\frac{N_1}{N_2}\sqrt{3}\,I\measuredangle -\varphi - 90\right)^* \tag{2.36}$$

$$= \left(1.5\,VI\measuredangle\varphi\right) + \left(1.5\,VI\measuredangle\varphi\right) = 3\,VI\measuredangle\varphi = S_{\text{in}}$$

2.10 In this question, since the zero component of the current is transferred from the secondary to the primary, there is no need to convert the components. Suppose:

$$\hat{V}_a = \frac{400\text{ V}}{\sqrt{3}}\measuredangle 0, \quad \hat{V}_b = \frac{400\text{ V}}{\sqrt{3}}\measuredangle -120°, \quad \hat{V}_c = \frac{400\text{ V}}{\sqrt{3}}\measuredangle +120°$$

Voltages and turns of coils have a direct relationship, we have:

$$\hat{V}_{AB} = 20\text{ kV}\measuredangle 0, \quad \hat{V}_{BC} = 20\text{ kV}\measuredangle -120°, \quad \hat{V}_{CA} = 20\text{ kV}\measuredangle +120°$$

Note that the ratio of turns of the coils is obtained from the rated voltages:

$$\frac{\hat{V}_{AB}}{\hat{V}_a} = \frac{20\text{ kV}\measuredangle 0}{\frac{400\text{ V}}{\sqrt{3}}\measuredangle 0} = 50\sqrt{3} = \frac{N_1}{N_2}$$

Then:

$$\hat{V}_A = \frac{20\text{ kV}}{\sqrt{3}}\measuredangle -30°, \quad \hat{V}_B = \frac{20\text{ kV}}{\sqrt{3}}\measuredangle -150°, \quad \hat{V}_C = \frac{20\text{ kV}}{\sqrt{3}}\measuredangle +90°$$

Power equations in loads and calculation of transformer secondary current:

$$\text{Load } A: \hat{I}_a = \frac{5\text{ kW}}{\frac{400\text{ V}}{\sqrt{3}}(0.5)}\measuredangle(0 - \cos^{-1}0.5) = 43.3\ A\measuredangle -60°$$

$$\text{Load } B: \hat{I}_b = \frac{8\text{ kW}}{\frac{400\text{ V}}{\sqrt{3}}(0.5)} \measuredangle(-120-\cos^{-1}0.5) = 69.28\, A\measuredangle -180°$$

$$\text{Load } C: \hat{I}_c = 0$$

$$\hat{I}_n = \hat{I}_a + \hat{I}_b + \hat{I}_c = 43.3\, A\measuredangle -60° + 69.28\, A\measuredangle -180° = 60.62\, A\measuredangle -141.8°$$

Primary winding current (inverse of the conversion ratio):

$$\hat{I}_{AB} = \frac{1}{50\sqrt{3}} 43.3\, A\measuredangle -60° = 0.5\, A\measuredangle -60°$$

$$\hat{I}_{BC} = \frac{1}{50\sqrt{3}} 69.28\, A\measuredangle -180° = 0.8\, A\measuredangle -180°$$

$$\hat{I}_{CA} = 0$$

KCL in the primary transformer:

$$\begin{cases} \hat{I}_A = \hat{I}_{AB} - \hat{I}_{CA} = 0.5\, A\measuredangle -60° \\ \hat{I}_B = \hat{I}_{BC} - \hat{I}_{AB} = 0.8\, A\measuredangle -180° - 0.5\, A\measuredangle -60° \\ \hat{I}_C = \hat{I}_{CA} - \hat{I}_{BC} = 0 - 0.8\, A\measuredangle -180° \end{cases} \Rightarrow \begin{cases} \hat{I}_A = 0.5\, A\measuredangle -60° \\ \hat{I}_B = 1.136\, A\measuredangle 157.6° \\ \hat{I}_C = 0.8\, A\measuredangle 0° \end{cases} \Rightarrow$$

Now we check the output and input power to the transformer:

$$S_{\text{out}} = S_{\text{Load }A} + S_{\text{Load }B} = \left(5\text{ kW} + j5\text{ k}\tan\left(\cos^{-1}0.5\right)\right) + \left(8\text{ kW} + j8\text{ k}\tan\left(\cos^{-1}0.5\right)\right)$$

$$S_{\text{out}} = (5\text{ kW} + j8.66\text{ kVAr}) + (8\text{ kW} + j13.856\text{ kVAr}) = 13\text{ kW} + j22.516\text{ kVAr}$$

$$S_{\text{in}} = \hat{V}_A\hat{I}_A^* + \hat{V}_B\hat{I}_B^* + \hat{V}_C\hat{I}_C^* \Rightarrow$$

$$S_{in} = \frac{20\text{ kV}}{\sqrt{3}}\left(0.5\, A\measuredangle(-30+60)° + 1.136\, A\measuredangle(-150-157.6)° + 0.8\, A\measuredangle(90+0)°\right) \Rightarrow$$

$$S_{\text{in}} = 13.004\text{ kW} + j\,22.517\text{ kVAr} \approx S_{\text{out}}$$

2.11 From equation (2.20), we have:

$$\overbrace{\begin{bmatrix} \hat{I}_a^0 \\ \hat{I}_a^+ \\ \hat{I}_a^- \end{bmatrix}}^{I_a^{0+-}} = \overbrace{\frac{1}{3}\begin{bmatrix} 1 & 1 & 1 \\ 1 & \alpha & \alpha^2 \\ 1 & \alpha^2 & \alpha \end{bmatrix}}^{A^{-1}} \overbrace{\begin{bmatrix} \hat{I} \\ 0 \\ 0 \end{bmatrix}}^{I^{abc}} = \frac{\hat{I}}{3}\begin{bmatrix} 1 \\ 1 \\ 1 \end{bmatrix} \Rightarrow \hat{I}_a^0 = \hat{I}_a^+ = \hat{I}_a^- = \frac{\hat{I}}{3}$$

On the primary side, the positive and negative components are transferred in a ratio of one. However, the zero component is not transferred. Therefore, we have:

$$\hat{I}_A = \hat{I}_a^+ + \hat{I}_a^- = 2\left(\frac{\hat{I}}{3}\right)$$

The figures below show the sum of the effects of the components (Figures 2.37–2.40).

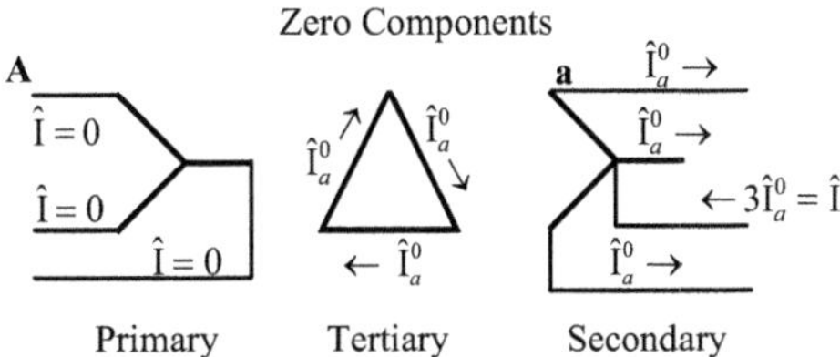

FIGURE 2.37 Zero components of three-winding transformer.

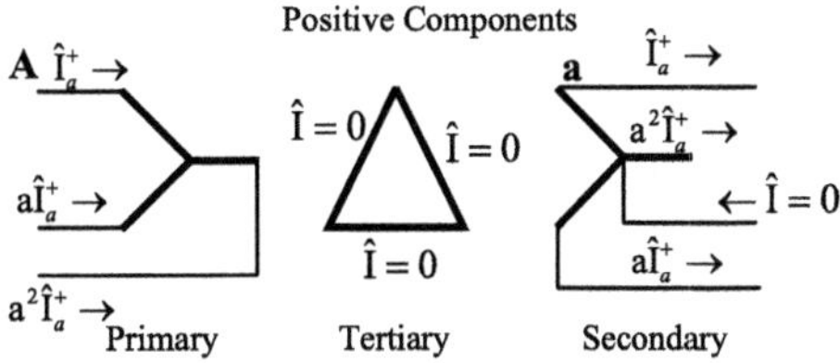

FIGURE 2.38 Positive components of three-winding transformer.

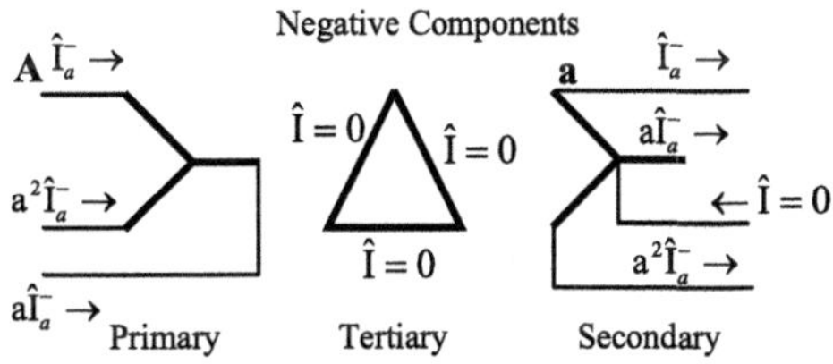

FIGURE 2.39 Negative components of three-winding transformer.

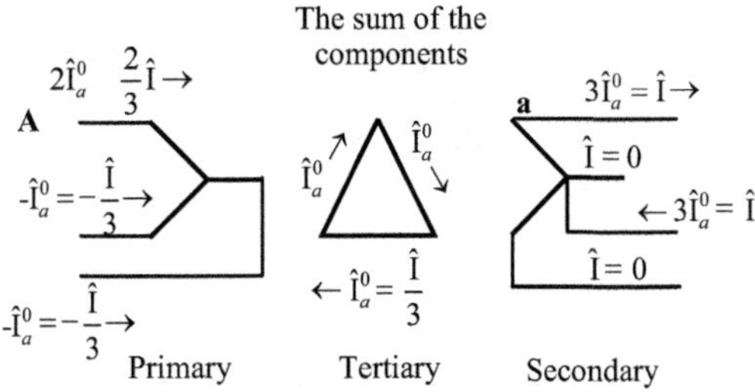

FIGURE 2.40 The sum of the effects of the components in three-winding transformer.

3 Special Transformers

Part One: Lesson Summary

3.1 INSTRUMENT TRANSFORMERS (CTs AND PTs)

Power systems deal with colossal voltages and currents. While these high values are necessary for efficient power transmission and distribution, directly measuring them can be dangerous, even impractical. This is where instrument transformers come in and safely reduce these high voltages and currents to more manageable levels for measuring and protection devices. There are two main types of instrument transformer:

Current transformers (CTs): These transformers reduce high currents to safer values for ammeters (ampere meter), relays, and other monitoring instruments. They work based on the principle of electromagnetic induction, where the primary current flowing through the CT's primary winding induces a proportional current in the secondary winding. The ratio of the number of turns in the primary to the number of turns in the secondary winding determines the current reduction factor.

Voltage transformers (VTs) [or potential transformers (PTs)]: Similar to CTs, VTs step down high voltages to safer levels for instruments like voltmeters and protective relays. They also operate based on the principle of electromagnetic induction, with the voltage ratio between the primary and secondary windings determined by the turn ratio.

Figures 3.1 and 3.2 show several CT and PT types. Note that the calculations for CT and PT are similar to the real transformer.

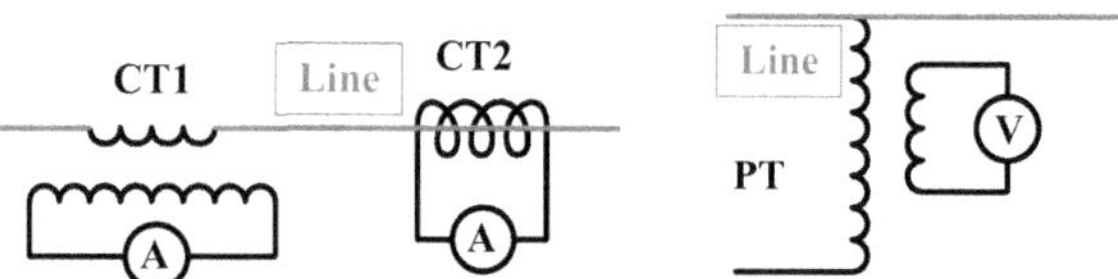

FIGURE 3.1 Symbolic figures of CT and PT.

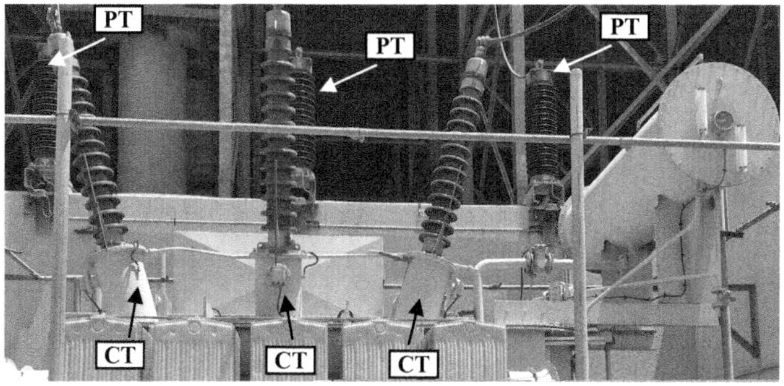

FIGURE 3.2 Real figures of CT and PT.

DOI: 10.1201/9781003537564-3

3.1.1 Benefits

- Safety: By reducing high voltages and currents, instrument transformers protect personnel and equipment from the hazards of direct contact with high-energy circuits.
- Accuracy: Instrument transformers allow for accurate measurement of high voltages and currents using standard measuring devices.
- Protection: They play a crucial role in protecting power system components by providing accurate data to protective relays. These relays can then detect faults and isolate them quickly, minimizing damage to equipment and preventing outages.
- Isolation: Instrument transformers provide galvanic isolation between the high-voltage side and the measurement or protection circuits. This isolation helps prevent ground faults and improves overall system stability.

3.1.2 Applications

- Power metering: They enable accurate measurement of electrical energy consumption in homes, industries, and commercial buildings.
- Protection systems: They provide data for protective relays that detect faults like short circuits and overloads, ensuring the safe operation of power systems.
- Monitoring and control: Instrument transformers provide vital information for system monitoring and control centers, allowing for real-time analysis of power flow and system health.
- Data acquisition: They play a role in data acquisition systems that collect information for power system analysis, optimization, and planning.

3.2 PHASE-SHIFTING TRANSFORMERS (PST)

A specialized type of transformer called a phase angle regulating transformer, also known as a phase angle regulator (PAR), phase-shifting transformer, phase shifter, or quadrature booster (quad booster), is used to regulate the active power flow in transmission networks without directly altering generation or consumption levels.

Unlike traditional transformers that simply change voltage levels, PSTs manipulate the phase angle between two points in the power grid, influencing how power flows across those lines.

A PST is a specialized transformer with two key components: a shunt transformer and a series transformer. The shunt transformer connects to the transmission line at a high impedance, while the series transformer is connected in series with the line at a very low impedance. By adjusting the tap position of the shunt transformer, PSTs control the voltage it injects into the line. This injected voltage has a specific phase angle relative to the existing line voltage. By varying the phase angle of the injected voltage, PSTs can influence the overall phase angle difference between the two connected systems.

The phase angle difference between the sending and receiving ends of a transmission line directly affects active power flow. By manipulating the phase angle with PSTs, we can influence how much power flows across the line, directing it to areas with higher demand or preventing the overloading of specific lines to perform congestion management.

3.2.1 Benefits

- Enhanced power flow control: PSTs offer a flexible and efficient way to manage power flow without changing generation or consumption patterns. This optimizes grid utilization and helps maintain system stability.
- Reduced transmission losses: By optimizing power flow and avoiding overloads, PSTs can help minimize energy losses during transmission, leading to a more efficient power grid.

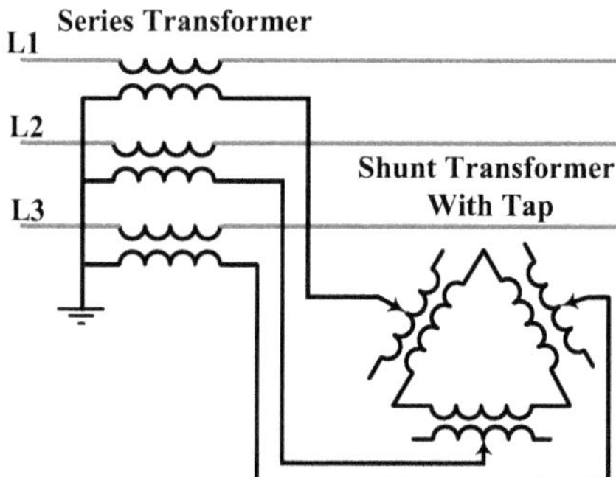

FIGURE 3.3 Equivalent circuit of the PST.

- Improved system reliability: Precise power flow control with PSTs allows for better management of contingencies such as line outages or sudden changes in demand. This enhances overall system reliability and reduces the risk of blackouts.
- Integration of renewables: With the increasing penetration of renewable energy sources such as wind and solar, PSTs play a crucial role in managing the variable power output from these sources, ensuring grid stability.

The equivalent circuit of the PST can be seen in Figure 3.3.

3.3 HVDC TRANSFORMERS (HVDC POWER CONVERTER TRANSFORMER)

The world of power transmission is constantly evolving, and High-Voltage Direct Current (HVDC) transmission is playing an increasingly important role. HVDC transformers act as the missing link between HVAC and HVDC systems. HVDC transformers facilitate the conversion of AC voltage from the grid into a controllable DC voltage for the HVDC system. This conversion happens through a complex process involving multiple windings, power electronic converters, and sophisticated control systems. They also handle the conversion back from DC to AC voltage when feeding power back into the AC grid at the receiving end. This ensures compatibility with existing infrastructure. Moreover, HVDC transformers provide isolation between the AC and DC sides. This isolation minimizes leakage currents and improves overall system stability [31].

3.3.1 Benefits

- Efficient long-distance transmission: HVDC transmission offers lower energy losses compared to AC, making it ideal for transmitting power over long distances. HVDC transformers facilitate this by enabling efficient conversion between AC and DC.
- Improved grid integration: HVDC transformers allow for the seamless integration of renewable energy sources with variable power output into the grid. They handle the conversion between AC from renewable sources and DC for HVDC transmission [32–42].
- Enhanced power flow control: HVDC systems offer greater control over power flow compared to AC. HVDC transformers enable this control by providing a conversion point where the DC voltage can be regulated, influencing power flow across the transmission line.
- Reduced environmental impact: With lower energy losses during transmission, HVDC contributes to a more environmentally friendly power grid.

Figure 3.4 shows a typical HVDC converter transformer.

FIGURE 3.4 A typical HVDC converter transformer.

3.4 DISTRIBUTION TRANSFORMERS

Distribution transformers play a crucial role in stepping down high-voltage electricity to safe and manageable levels for everyday use. These transformers are the final link in the power delivery chain. They receive high-voltage electricity (typically ranging from 11 to 33 kV) from transmission lines and transform it into lower voltages (usually 240 or 380 V) suitable for powering homes, businesses, and industries.

These transformers are relatively simple in design, consisting of a core made from laminated steel and two sets of windings – primary and secondary. The high-voltage AC current passes through the primary winding, generating a magnetic field in the core. This magnetic field then induces a voltage in the secondary winding but at a lower voltage based on the ratio of the number of turns in each winding.

Distribution transformers come in various types, including pole-mounted transformers for residential areas, and pad-mounted transformers for commercial and industrial applications. In addition, the modular design of distribution transformers allows for easy adaptation to varying power demands in different areas. In the following, you can see an example of rated power of distribution transformers:

- Distribution transformers fixed to wooden or concrete poles (pole-mounted transformers): 25, 50, 75, and 100 kVA
- Distribution transformers mounted on two concrete or wooden poles: 125, 160, 200, 250, 315, and 400 kVA
- A pad-mount (or pad-mounted transformer) is a ground-mounted electric power distribution transformer in a locked steel cabinet mounted on a concrete pad. 500, 630, 800, 1,000, 1,250, 1,600, 2,000, and 2,500 kVA

Figure 3.5 shows a variety of real examples of distribution transformers.

FIGURE 3.5 A variety of real examples of distribution transformers.

3.5 ELECTRICAL ARC FURNACE (EAF) TRANSFORMERS

For the purpose of providing the required power at a low voltage level, Electric Arc Furnaces (EAF), Ladle Furnaces (LF), high current rectifiers, and converter transformers require a specific design. Electrical arc furnace transformers, also known as melting transformers, are the main element behind EAFs used in steel production and other industrial processes. Unlike regular transformers, they produce very high currents at different voltage levels.

EAF transformers take high voltage, low current electricity from the grid and convert it to low voltage, incredibly high current suitable. While secondary, currents exceeding 100 kA and power ratings ranging from 10 to 300 MVA are typical.

These transformers are built tough to withstand demanding electrical conditions. Frequent current fluctuations, unbalanced loads, and scorching temperatures are all part of the job description. Special design features ensure they can handle it all. EAF operation involves melting periods requiring a significant power boost. To cope with this, these transformers are built to deliver up to 20% overload capacity for short durations.

Electric arc furnace transformers can be divided into two groups, AC and DC.

EAF transformers play a pivotal role in various industries. They are the backbone of steel production. Melting scrap steel in electric arc furnaces is a crucial step in steel manufacturing. In addition, the production of certain metals like white corundum and electrolytic aluminum also relies on EAFs powered by these transformers. Figure 3.6 shows the equivalent circuit and a real example of electrical arc furnace transformers.

3.6 ISOLATION TRANSFORMERS

Isolation transformers are electrical guardians, protecting equipment from the perils of power. Unlike regular transformers that simply change voltage levels, isolation transformers provide a crucial layer of separation between the incoming power source and the device being powered. Usually, the isolated transformer has a 1:1 turn ratio.

3.6.1 Benefits

- Shock prevention: By creating a physical barrier between the primary and secondary circuits, isolation transformers significantly reduce the risk of electric shock. This is particularly important when working with sensitive electronics or in environments with potential grounding issues.
- Ground loop elimination: Ground loops arise from minute voltage differences between equipment grounds. Isolation transformers break this loop, preventing unwanted current flow and potential damage to grid equipment.

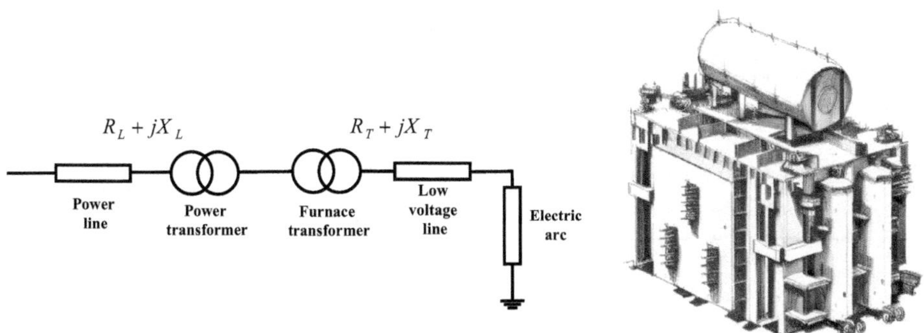

FIGURE 3.6 The equivalent circuit and a real example of electrical arc furnace (EAF) transformers.

- Noise reduction: Power lines can carry electrical noise that interferes with the proper functioning of sensitive equipment. Isolation transformers act as filters, attenuating this noise and ensuring a cleaner, more stable power supply for devices.
- Signal integrity: In applications involving delicate signals, even minor fluctuations can be detrimental. Isolation transformers minimize these disruptions, safeguarding the integrity of the signals passing through.

3.6.2 Applications

- Medical equipment: Patient safety is paramount in medical settings. Isolation transformers are employed to safeguard medical devices from potential leakage currents and ensure reliable operation.
- Electronic components: Sensitive electronic components used in computers, audio/video equipment, and communication systems all benefit from the clean power and noise reduction provided by isolation transformers.
- Industrial machinery: Isolation transformers help protect industrial control systems and programmable logic controllers (PLCs) from power surges and electrical noise, ensuring the smooth operation of automated processes.

3.6.3 Choosing the Right

- Power requirements: The transformer's capacity should meet the power needs of the device.
- Input/output voltage: Ensure the transformer handles the incoming voltage and delivers the desired output voltage for the equipment.
- Environmental conditions: Consider factors like temperature and humidity when choosing a suitable transformer.

Figure 3.7 shows a real example and the main reason for using isolation transformers.

3.7 DRY-TYPE TRANSFORMERS

Dry-type transformers are a safe and versatile alternative to their liquid-filled counterparts. Unlike traditional transformers that use oil for cooling and insulation, dry-type transformers utilize air, inert gas, or solid insulating materials. This design offers several advantages, making them suitable for various applications.

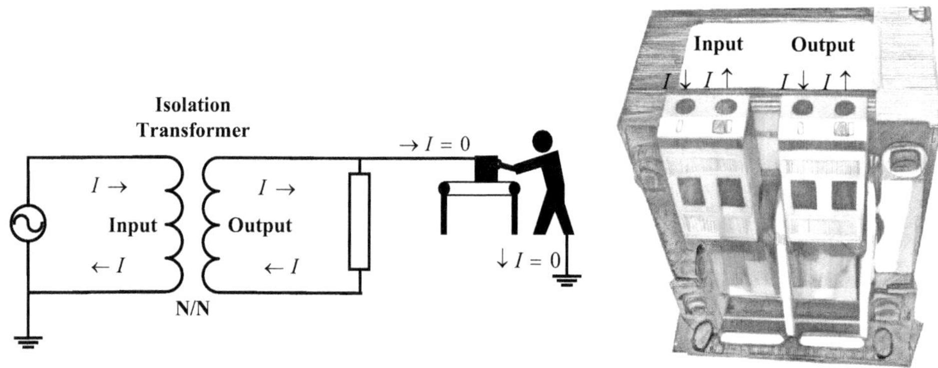

FIGURE 3.7 A real example and the main reason for using isolation transformers.

3.7.1 Advantages

- Fire safety: Eliminating flammable liquids significantly reduces the risk of fire hazards associated with transformer malfunctions. This makes dry-type transformers ideal for use in buildings, commercial spaces, and areas with fire safety concerns.
- Environmental benefits: In case of leaks or accidents, dry-type transformers pose minimal environmental hazards compared to oil-filled transformers, which can contaminate soil and water.
- Cleanliness: Dry-type transformers do not require maintenance associated with oil changes and leak inspections, leading to lower maintenance costs.
- Wider installation flexibility: Since they do not pose a fire risk, dry-type transformers can be installed in places where oil-filled transformers are restricted, like buildings with tight spaces or close to sensitive equipment.

3.7.2 Types

- Ventilated: These transformers rely on natural air circulation through vents for cooling. They are suitable for environments with decent air circulation and low power requirements.
- Sealed: These transformers utilize air circulation achieved by fans or natural convection within the enclosure. They are well suited for dusty or dirty environments.
- Cast resin: These transformers have their windings encased in fire-retardant epoxy resin. They offer excellent fire resistance and are compact, making them ideal for space-constrained applications.

3.7.3 Applications

- Buildings: Dry-type transformers are widely used in commercial buildings, offices, and high-rise constructions due to their safety and cleanliness.
- Manufacturing facilities: In industrial settings, they provide a reliable power source for machinery and control systems, especially in areas with potential flammability concerns.
- Renewable energy integration: They play a crucial role in integrating renewable energy sources like solar and wind power into the grid.

3.7.4 Choosing the Right

- Power needs: The transformer's capacity should match the power requirements.
- Environmental conditions: Choose a transformer suitable for the ambient temperature, humidity, and dust levels of the installation site.
- Noise levels: Some dry-type transformers, particularly ventilated types, can generate noise. Consider noise limitations if the transformer will be installed in a noise-sensitive area.

Figure 3.8 shows a real example of a dry-type transformer.

3.8 WELDING TRANSFORMERS

Welding transformers are the main element behind arc welding processes, the most common type of welding. They perform a vital task: transforming high-voltage, low-current utility power into low-voltage, high-current electricity ideal for creating a welding arc. Incoming AC power from the wall socket (typically 220 or 380 V) boasts high voltage but relatively low current. Welding transformers dramatically reduce the voltage and significantly increase the current. This high current is what enables the formation and maintenance of the intense heat required for welding. Many

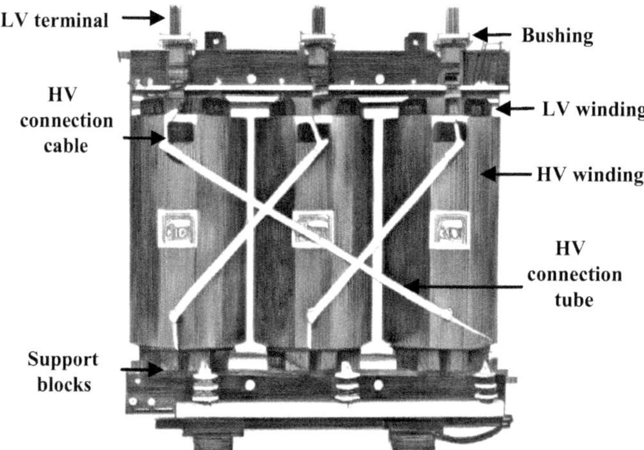

FIGURE 3.8 A real example of a dry-type transformer.

welding transformers come equipped with tap changers or variable controls. These allow the welder to fine-tune the output current to match the thickness of the metal being welded and the specific welding process being used.

3.8.1 Types

- AC transformers: The most common type, AC transformers deliver alternating current (AC) for welding. They are known for their affordability and durability.
- DC transformers are used for special welding processes that require more control and penetration. DC transformers provide direct current (DC) output. DC transformers are made by combining conventional transformers with DC–DC converters.
- Inverter-based transformers: A modern advancement, inverter-based transformers are lighter, more energy-efficient, and offer superior arc control compared to traditional transformers.

3.8.2 Key Features

- Duty cycle: This rating specifies the percentage of a given period that the transformer can operate at its rated output current without overheating. It is crucial to choose a transformer with a duty cycle exceeding the typical welding usage.
- Electrode compatibility: Different transformers are designed to work with specific electrode types. Ensure the transformer is compatible with the electrodes intended to be used.
- Portability: For projects requiring mobility, a lightweight, portable transformer is preferred. Conversely, larger projects might necessitate a more powerful, stationary transformer.

Figure 3.9 shows the equivalent circuit and a real example of welding transformers.

3.8.3 Applications

- Arc welding: This term covers various processes that utilize the heat from an electric arc to fuse metals. Common applications include metal fabrication, repair work, and construction.
- Cutting: Specialized welding transformers can produce a high-temperature plasma arc for cutting through metals.

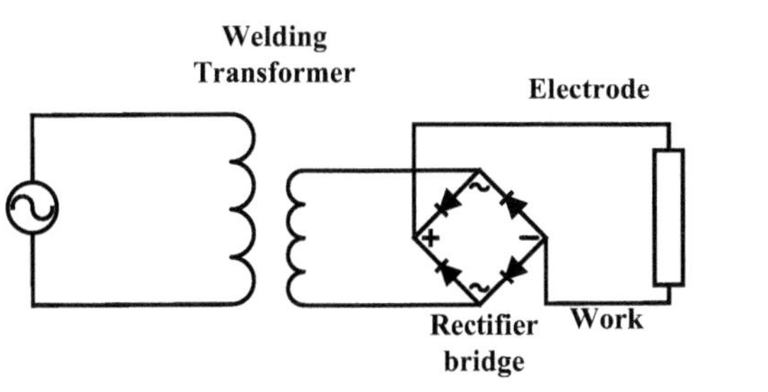

FIGURE 3.9 The equivalent circuit and a real example of welding transformers.

3.9 AUDIO-FREQUENCY (AF) TRANSFORMERS

Audio frequency transformers (AF) are not typically used in power systems. AF transformers are designed for the range of human hearing, around 20 Hz to 20 kHz. However, there can be some niche applications of AF transformers in conjunction with power systems for purposes like control, signaling, and leakage detection. In some cases, audio frequencies might be used to identify leaks in power system components through special transformers.

3.9.1 Important Points

- Impedance matching: Audio equipment often has mismatched impedances, which can lead to weak or distorted signals. AF transformers act as bridges, adjusting the impedance of the input signal to match the output device, ensuring maximum power transfer and optimal signal quality.
- Signal balancing: In balanced audio systems, unwanted noise and hum are minimized by using balanced cables with two identical conductors carrying the signal and its inverted version. Audio transformers can convert unbalanced signals (single conductor with ground) to balanced signals and vice versa, promoting cleaner audio transmission.
- Signal isolation: Transformers provide electrical isolation between circuits, preventing ground loops and potential voltage spikes from damaging sensitive audio equipment. This isolation also helps eliminate unwanted hum and noise that can creep into the audio signal.

3.9.2 Applications

- Microphone preamps: Microphones often have high-impedance outputs. AF transformers can boost the weak microphone signal and match its impedance for better compatibility with microphone preamps.
- Output transformers: In some vintage tube amplifiers, output transformers match the high impedance of the output stage to the lower impedance of loudspeakers, ensuring efficient power delivery.
- Audio signal coupling: Transformers can be used to couple different stages in an audio circuit, like preamps to power amps, while blocking DC voltage and allowing only the desired AC audio signal to pass through.

3.9.3 Choosing the Right

- Frequency response: The transformer should have a frequency response that covers the entire audible range (20 Hz to 20 kHz) with minimal distortion.

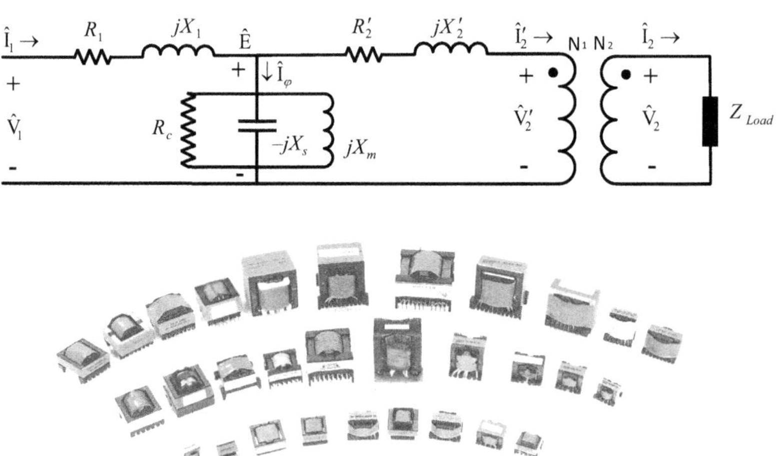

FIGURE 3.10 The equivalent circuit and real examples of AF transformers.

- Impedance ratio: Choose a transformer with an impedance ratio that matches the input and output impedances of the devices.
- Power rating: The transformer's power rating should be sufficient to control the audio power levels used.

Figure 3.10 shows the equivalent circuit and real examples of AF transformers.

3.10 GROUNDING TRANSFORMERS

Grounding transformers are one of the most important members of electrical safety and stability in power systems. Unlike regular transformers that convert voltage levels, grounding transformers focus on a low-impedance path to ground for ungrounded electrical systems.

3.10.1 Importance

- Safety: Grounding transformers play a vital role in ensuring the safety of personnel and equipment. They provide a defined path for fault currents to flow in case of an electrical fault, preventing dangerous voltage surges on equipment enclosures. This helps minimize the risk of electrical shock and equipment damage.
- System stability: Grounding transformers also contribute to the overall stability of the power grid. By providing a controlled path for fault currents, they help prevent uncontrolled current flow that can lead to voltage fluctuations or even blackouts.

3.10.2 Types

- Zigzag transformer: This is the most common type. It utilizes a single winding with a special zigzag connection pattern to create a path for current to flow to the ground.
- Delta-wye transformer: This type employs a transformer with a delta-connected primary winding and a wye-connected secondary winding with a grounded neutral point.

3.10.3 Applications

- Wind farms: In large wind farms, grounding transformers are crucial for safety and fault protection on ungrounded turbine strings.

- Power plants: Neutral grounding transformers are a common element in power plants, providing a grounding path for generators.
- High-voltage transmission systems: Grounding transformers can be used at substations to ground high-voltage lines that might operate ungrounded.

3.10.4 Benefits

- Enhanced safety: This transformer can reduce the risk of electric shock and protect personnel and equipment.
- Improved system stability: A controlled path for fault currents helps maintain grid stability and prevent cascading outages.
- Reduced transient voltages: Grounding transformers can help limit transient voltage spikes during fault conditions.

3.10.5 Selection Considerations

- System voltage and current: The transformer's capacity should match the voltage and current ratings of the system it protects.
- Grounding impedance: The transformer's impedance characteristic determines how effectively it channels fault currents to ground.
- System configuration: The specific type of grounding transformer (zigzag or delta-wye) is chosen based on the configuration of the power system.

Figure 3.11 shows the equivalent circuit and a real example of grounding transformers.

3.11 ROTARY TRANSFORMERS

Rotary transformers are specialized transformers designed to transfer electrical power between two parts that rotate relative to each other. Unlike stationary transformers that use a fixed magnetic core, rotary transformers employ a unique design to achieve this feat.

Rotary transformers come in two main shapes: cylindrical for transmitting higher power and pancake-shaped for lower power applications. The core of a rotary transformer is split into two halves, a primary half mounted on the stationary part and a secondary half fixed to the rotating part. These halves are separated by a small air gap.

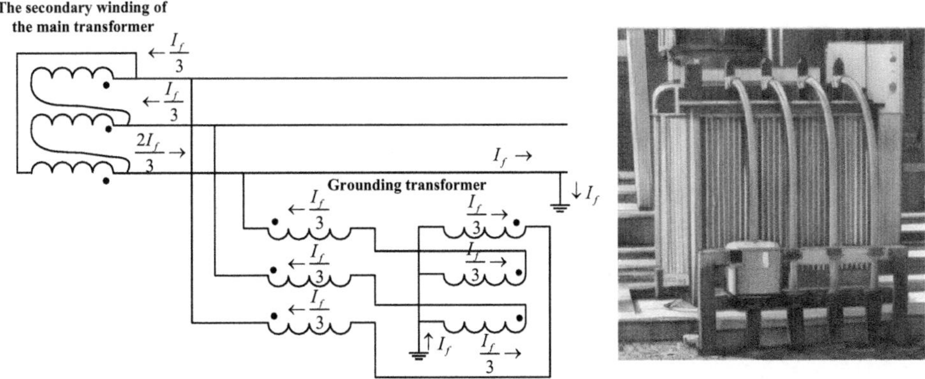

FIGURE 3.11 The equivalent circuit and real example of grounding transformers.

3.11.1 Applications

- Wind turbines: These transformers are crucial for transferring electrical power generated by the rotating turbine shaft to the stationary grid connection.
- Robotic arms and automation: Rotary transformers enable continuous power and signal flow between the base and the rotating joints of robots, which is vital for their operation.
- Radars and communication systems: In rotating radar antennas or communication equipment, rotary transformers ensure uninterrupted signal transmission.
- Medical equipment: Some medical equipment, like CT scanners or centrifuges, utilize rotary transformers for power or signal transfer within their rotating components.

3.11.2 Advantages

- Continuous power transfer: Unlike solutions using brushes or slip rings, which can wear out and create electrical noise, rotary transformers provide a smooth and reliable path for uninterrupted power or signal flow.
- Sealed design: The enclosed design protects the transformer from dust, moisture, and contaminants, making it suitable for harsh environments.
- Low maintenance: With no physical contact between the rotating and stationary parts, rotary transformers require minimal maintenance compared to brush-based solutions.

3.11.3 Selection Considerations

- Voltage and current requirements: The transformer's capacity, which should match the power, needs of the application.
- Speed of rotation: The maximum rotational speed, which the transformer is designed to handle.
- Environmental conditions: Choose a transformer suitable for the temperature, humidity, and any potential contaminants in the operating environment.

Figure 3.12 shows the equivalent circuit and real examples of rotary transformers.

3.12 VARIABLE FREQUENCY TRANSFORMER (VFT)

Variable frequency transformers (VFTs) are a relatively new innovation in the world of power transmission. Unlike traditional transformers that simply adjust voltage levels, VFTs tackle a more complex challenge: enabling power exchange between asynchronous and alternating current (AC) grids. The frequency can vary slightly between different power grids. This inconsistency creates a barrier

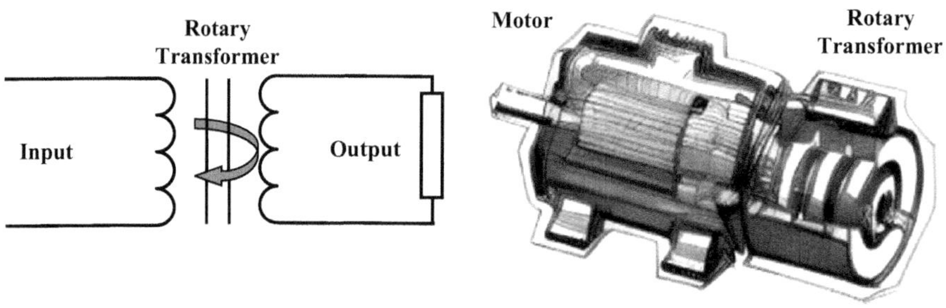

FIGURE 3.12 The equivalent circuit and real examples of rotary transformers.

to directly connecting these grids and sharing power. VFTs act as intermediaries, allowing power to flow between asynchronous grids. They achieve this through a combination of technologies:

- Doubly fed electric machine: At its core, a VFT resembles a vertical shaft hydroelectric generator with a three-phase wound rotor.
- Power converter: This component converts the incoming AC power from one grid into a variable frequency AC.
- Rotary transformer: This transformer transfers the power from the converter to the other grid, while adjusting the frequency to match the recipient grid's requirements.

3.12.1 Benefits

- Enhanced grid integration: VFTs facilitate the connection of previously incompatible grids, enabling a more interconnected and flexible power grid.
- Improved power flow management: They allow for better control of power flow between grids, optimizing resource utilization and grid stability.
- Integration of renewable energy: VFTs can play a crucial role in integrating renewable energy sources like wind and solar power into the grid, which often have variable output.

3.12.2 Applications

- Interconnecting national grids: VFTs hold promise for connecting national grids with slightly different AC frequencies, facilitating regional power exchange.
- Connecting offshore wind farms: As offshore wind farms become more prevalent, VFTs can be instrumental in efficiently integrating their power into the mainland grid.
- Microgrid integration: VFTs can be used to manage power flow within microgrids, which are localized grids combining various power sources and loads.

3.12.3 VFT Implementation

- Cost: VFTs are currently a complex and expensive technology. As the technology matures, the cost is expected to decrease.
- System complexity: Integrating VFTs into existing grids requires careful planning and system upgrades to ensure smooth operation.

Figure 3.13 shows the equivalent circuit and a real example of VFT.

3.13 PLANAR TRANSFORMERS

Planar transformers are a distinct category of transformers known for their low profile, flat design compared to traditional coil-wound transformers. This unique design offers several advantages, making them well-suited to various applications requiring efficient power conversion in space-constrained environments.

3.13.1 New Structure

- Flat: Unlike conventional transformers with bulky windings wrapped around a bobbin, planar transformers utilize flat, printed circuit board (PCB) technology. The primary and secondary windings are etched or laminated onto separate PCB layers, creating a much flatter profile.

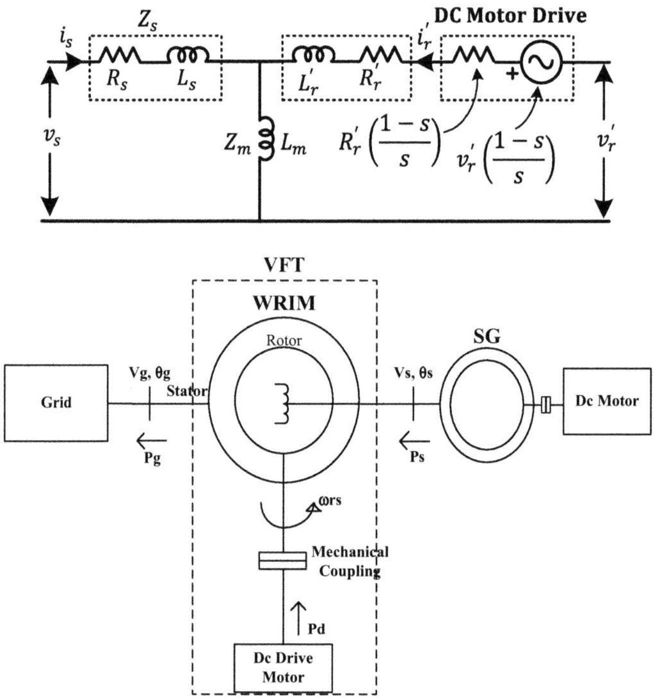

FIGURE 3.13 The equivalent circuit and a real example of VFT.

- Material: Planar transformers often use thin sheets of magnetic material like ferrite or amorphous metal for their cores. These materials offer high efficiency at high frequencies, making them ideal for applications involving fast-switching power supplies.

3.13.2 Advantages

- Compact design: The flat profile allows for significant space savings compared to traditional transformers, making them valuable in applications where space is at a premium, like portable electronics and compact power supplies.
- High-efficiency: The planar construction allows for tight coupling between the primary and secondary windings, minimizing leakage inductance and core losses, leading to improved efficiency.
- High-frequency operation: Due to the use of thin magnetic materials and distributed windings, planar transformers can operate efficiently at higher frequencies compared to their bulkier counterparts.
- Lower EMI: The planar design can help reduce electromagnetic interference (EMI) compared to traditional transformers, minimizing unwanted noise in sensitive electronic circuits.

3.13.3 Applications

- Switched-mode power supplies (SMPS): Planar transformers are prevalent in SMPS used in computers, laptops, and various electronic devices due to their compact size and high switching frequency capability.

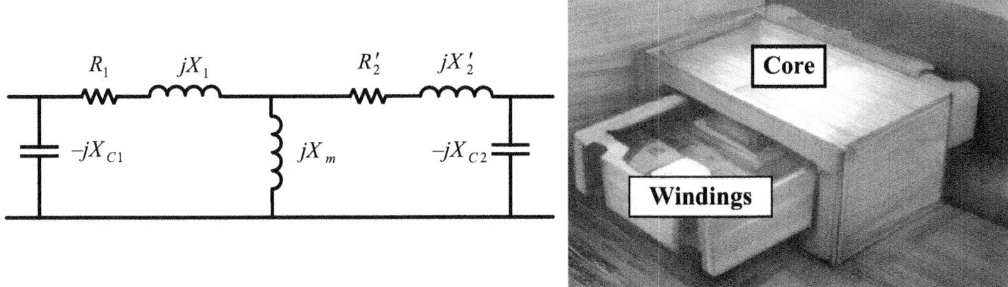

FIGURE 3.14 The equivalent circuit and real examples of a planar transformer.

- DC–DC converters: These converters are essential for regulating power within electronic devices, and planar transformers offer a space-saving and efficient solution for this application.
- High-speed data transmission: In data communication systems, planar transformers can be used for signal isolation and impedance matching at high frequencies.
- Wireless power transfer: Planar transformers hold promise for applications like wireless charging due to their ability to efficiently transfer power over short distances.

3.13.4 Selection Considerations

- Power requirements: The transformer's capacity should match the voltage and current needs of your application.
- Operating frequency: Consider the frequency range at which the transformer will operate.
- Inductance and impedance: Choose a transformer with the desired inductance and impedance values for your circuit.
- Environmental conditions: Consider factors like temperature and humidity when selecting a suitable transformer.

Figure 3.14 shows the equivalent circuit and real examples of a planar transformer.

Part Two: Answer-Question

3.14 TWO-CHOICE QUESTIONS (YES/NO)

1. The calculations CT and PT are similar to the real transformer.
2. CTs and PTs are instrument transformers.
3. The secondary of the CT can be an open circuit similar to the PT.
4. PST and PAR is a specialized form of transformer used to control the flow of active power.
5. The shunt transformer connects to the transmission line at a low impedance.
6. PST helps maintain system stability.
7. PST enhances overall system reliability and reduces the risk of blackouts.
8. HVDC transmission offers more energy losses compared to AC.
9. HVDC transmission makes it ideal for transmitting power over short distances.
10. HVDC systems offer less control over power flow compared to AC.
11. HVDC has higher energy losses during transmission.
12. The transformer with a power of 25 kVA is a type of pole-mounted transformer.
13. The transformer with a power of 100 kVA is a type of pad-mount or pad-mounted transformer.

14. Electric Arc Furnaces (EAF) need a specific design to supply the necessary power at a low current level.
15. EAF transformers take high voltage, low current electricity from the grid.
16. In EAF normal operation there are frequent current fluctuations and unbalanced loads.
17. Usually, the isolated transformer has a 1:1 turns ratio.
18. The ground loop is usually not eliminated by using an isolation transformer.
19. When choosing the right transformer, factors such as temperature and humidity are not important.
20. Dry-type transformers reduce the risk of fire hazards associated with transformer malfunctions.
21. Dry-type transformers play a crucial role in integrating renewable energy sources into the grid.
22. Some dry-type transformers can generate noise.
23. Welding transformers transform high-voltage, low-current utility power into low-voltage, high-current electricity.
24. AF transformers focus on the audio frequencies audible to the human ear.
25. Grounding transformer focuses on creating a high impedance path to ground for ungrounded electrical systems.
26. Grounding transformers also contribute to the overall stability of the power grid.
27. The specific type of grounding transformer (delta–delta) is chosen based on the configuration of the power system.
28. Rotary transformers come in pancake-shaped for transmitting higher power applications.
29. Some medical equipment utilizes rotary transformers.
30. Rotary transformers provide a smooth and reliable path for uninterrupted power or signal flow.
31. The rotating transformer is very sensitive to dust, moisture, and contaminants.
32. Variable frequency transformers (VFTs) enable power exchange between synchronous AC grids.
33. Rotary transformer is a type of VFT.
34. VFTs hold promise for connecting national grids with slightly different AC frequencies.
35. VFTs are currently a simple and inexpensive technology.
36. Planar transformers utilize flat, printed circuit board (PCB) technology.
37. Planar transformers hold promise for applications like wireless charging due to their ability to efficiently transfer power over short distances.
38. Planar transformers have low efficiency.

3.15 KEY ANSWERS TO TWO-CHOICE QUESTIONS

Yes	1,2,4,6,7,12,15,16,17,20,21,22,23,24,26,29,30,33,34,36,37
No	3,5,8,9,10,11,13,14,18,19,25,27,28,31,32,35,38

4 Transformer Control

Part One: Lesson Summary

4.1 VOLTAGE REGULATION

Maintaining the output voltage of the transformer is one of its important capabilities, whose changes are defined by voltage regulation. The output voltage of the ideal transformer is constant (from no-load to full-load), and in fact, it is the resistances and reactances of the real transformer that cause the output voltage to change while the load change. Even though ideal transformers have zero regulation, but in practice, values between 1% and 3% are often deemed optimal. In general, a lower voltage regulation percentage indicates a better ability to maintain a constant voltage.

The importance of voltage regulation percentage in equipment protection, load stability, and power grid efficiency is clearly seen. Sensitive electronic devices can be disrupted or damaged by significant voltage fluctuations. Constant voltage ensures their correct operation. In addition, electrical appliances and motors rely on a certain voltage range for effective operation, and voltage regulation minimizes performance issues caused by voltage drops or fluctuations. Finally, it can be said that throughout the transmission and distribution network, voltage regulation helps to minimize power losses and ensures reliable delivery of power to consumers.

Resistance, leakage reactance, and load current are effective factors in voltage regulation. The resistance of the transformer windings causes a voltage drop when current passes through them. Leakage flux results in a slight voltage difference between the ideal and actual output. Finally, as the load current on the secondary winding increases, the voltage drop caused by resistance becomes more significant.

In practice, no-load tap changers and on-load tap changers are used to improve the voltage regulation in transformers. No-load tap changer (NLTC), also known as off-circuit tap changer (OCTC) or de-energized tap changer (DETC). Tap changers allow adjustments to the number of turns on the primary winding, effectively altering the voltage ratio and compensating for voltage drops. Off-circuit tap changers require the transformer to be de-energized for adjustments. On-load tap changers as more sophisticated versions can adjust the tap ratio while the transformer is in operation.

4.2 TAP CHANGERS

A tap changer is a mechanism inside a transformer that allows the adjustment of the turns ratio in its windings, which in turn affects the voltage ratio between the input (primary) and output (secondary) sides. There are two main categories of tap changers based on performance: no-load tap changers (NLTC) and on-load tap changers (OLTC).

In NLTC or de-energized circuit tap changers (DETC), the transformer needs to be disconnected from the power source before adjustments can be made. NLTCs are simple and cheap, but not suitable for situations that require frequent voltage regulation. As a more advanced example, OLTCs can change tap positions while the transformer is energized. This ability enables constant voltage regulation, which is very important to maintain stability in power grids. OLTCs are more complex and expensive, but they also provide more control and flexibility.

Both NLTC and OLTC types use a series of connection points called taps in the transformer windings. These taps represent different parts of the coil with different numbers of turns. The

DOI: 10.1201/9781003537564-4

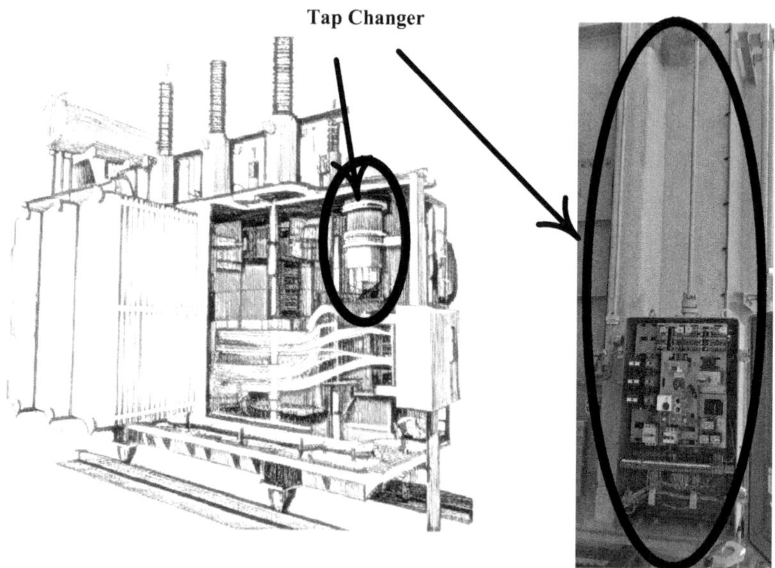

FIGURE 4.1 A tap changer in a power transformer.

tap-switching mechanism physically selects the appropriate tap on the primary winding, effectively changing the voltage ratio.

Tap changers are found in distribution, transmission, and industrial transformers. Distribution transformers step down high voltage from transmission lines to usable voltage levels for homes and businesses. Tap changers provide adjustments to compensate for voltage drops along the distribution line. Transmission transformers used in power plants and transmission lines handle high voltages. Tap changers help maintain grid stability by adjusting voltage over long line distances. In addition, many industrial processes require specific voltage levels, and tap changers enable adjustments to meet these specific needs.

Finally, by adjusting tap positions, we can ensure constant voltage delivery, protect equipment, and maintain the stability of power grids. Figure 4.1 shows the location of the tap changer in a power transformer. The control circuit and tap changer motor can be seen in the figure.

4.3 THERMAL CONSIDERATIONS

Transformers, like any kind of real machine, generate heat during operation, and thermal management is critical for ensuring the long life and reliable performance of the transformer. Sources of heat generation in transformers can be considered copper losses in windings, core losses, and stray losses. In fact, the primary source of heat in transformers is the electrical resistance of the windings that pass current through the windings. Energy is lost as heat due to this resistance. Also, the core, which is made of ferromagnetic materials, experiences energy losses during the magnetization process due to hysteresis and eddy currents, and these losses contribute to heat generation. The last is the leakage flux that leaks from the main magnetic field, creates eddy currents in nearby conductive components such as the tank and transformer shell, and adds to the overall heat of the transformer.

Excessive heat is harmful to the health of the transformer due to the destruction of the insulation, the reduction of efficiency, and the formation of hot spots.

Transformer insulation materials degrade faster at higher temperatures and this can lead to premature aging and possible failures. In fact, the rate of aging can approximately double with every 10-degree Celsius increase in temperature. In addition, with the increase in temperature, the electrical resistance of the coils also increases, which leads to more energy losses and a decrease in the overall efficiency of the transformer. On the other hand, it should be said that the uneven distribution of heat in the transformer could create local hot spots that destroy the insulation.

To deal with these thermal concerns, transformers use various cooling methods such as natural convection, forced air-cooling, natural oil circulation, and direct oil flow. In natural convection, air circulation around the transformer tank allows heat to dissipate naturally. This method is suitable for smaller transformers. In forced air-cooling, fans are used to circulate air more effectively, increasing heat transfer from the transformer core and windings to the surrounding air. In natural oil convection, many transformers use insulating oil as a coolant. The oil absorbs heat from the core and coils and circulates through the radiators where it releases heat to the surrounding air through natural convection. For larger transformers, pumps are used to pass the oil through the radiators, improving heat transfer efficiency.

Temperature sensors, load management, and regular maintenance can be used to monitor and maintain the thermal performance of transformers. Therefore, avoiding overloading the transformer helps to minimize heat generation. Periodic inspections and oil analyses ensure optimal cooling performance and identify potential thermal problems at an early stage.

Finally, by understanding thermal considerations and implementing effective cooling strategies, engineers can ensure that transformers operate within safe temperature ranges. This not only increases their lifespan but also ensures reliable and efficient power delivery.

4.4 TEMPERATURE CONTROL

The first step for effective temperature control is continuous monitoring, and transformers are equipped with temperature sensors placed inside the windings and oil. These sensors measure internal temperatures and can help operators identify potential overheating, optimize load management, and schedule maintenance. The most important use of temperature data is to determine maintenance intervals for cleaning cooling systems and replacing old oil, which optimizes cooling performance.

For critical applications or situations with high ambient temperature, complex control systems such as thermostats and on-load tap changers are used. Thermostats automatically activate cooling systems such as fans or pumps when the temperature exceeds a predetermined limit. In addition, in some transformers, on-load tap changers can be used to adjust the voltage ratio, which effectively reduces the current and thus heat generation.

Finally, transformer operators can maintain a safe thermal environment by implementing a combination of temperature monitoring, proper cooling methods, and advanced control systems. This not only ensures reliable power delivery but also extends the life of this critical grid equipment and minimizes maintenance costs and disruptions.

4.5 OVERLOADING EFFECTS AND PRECAUTIONS

As mentioned earlier, transformer overload causes heat increase and as a result, insulation destruction, efficiency reduction, core, and winding deformation, and finally shortens the life span of transformers.

Several measures can be taken to prevent transformer overload and ensure the safe and efficient operation of the transformer. Continuous load monitoring allows operators to identify situations where the load is approaching rated capacity and take corrective action before an overload occurs. Also, strategies such as reducing and temporarily cutting off non-critical loads or adding additional transformers can prevent overload. In addition, protective devices such as fuses or circuit breakers can disconnect the transformer from the power source in case of severe overload and prevent catastrophic failures. The final solution can be upgrading the transformer to a higher capacity.

4.6 REMOTE CONTROL AND MONITORING

In today's power networks, it is crucial to remotely control and monitor transformers. Remote control and monitoring technologies enable preventative maintenance for better efficiency and faster response times to potential issues. Remote control and monitoring increase efficiency, enable preventative maintenance, improve decision-making, reduce costs, and provide greater security. In fact, by analyzing data and identifying early warning signs, preventive maintenance can be planned and unexpected outages and costly repairs can be avoided. It also enables operators to make informed decisions about load management, tap changer settings, and resource allocation. This early detection of problems minimizes the risk of catastrophic failures, resulting in lower repair costs, reduced downtime, and greater personnel safety.

Sensors, a data acquisition unit (DAU), a communication gateway, and monitoring software are the main components of a remote monitoring system. The sensors collect data on temperature, oil level, pressure, vibration, and electrical characteristics and send it to the DAU. DAU converts this raw data into a digital format suitable for transmission. The communication gateway transmits this processed data to a central monitoring station using appropriate communication methods such as cellular networks, satellite communications, or power transmission line communications (PLCs). Finally, with the aid of appropriate software, the supervisory center provides an overview of the health and performance of the transformer to the operators. It is noteworthy that in some cases remote control functions may also be integrated into this system to enable limited remote control of the transformer.

Ultimately, it is important to emphasize that remote access significantly heightens the importance of cybersecurity measures. Therefore, it is very important to implement secure communication protocols, user authentication, and encryption techniques to protect the power network against cyberattacks and unauthorized access.

4.7 TRANSFORMER COOLING

The main matters related to transformer cooling tools have been mentioned in the previous sections. However, it is worth mentioning a few more details here. The choice of cooling method depends on the size, capacity, and application of the transformer. The methods are:

Natural convection (NC): This is the simplest method, suitable for smaller transformers that rely on natural air circulation around the transformer tank to dissipate heat. This method is economical but limits the capacity of the transformer.

Oil natural air natural (ONAN): In this method, oil is used to transfer internal heat. The oil absorbs heat from the core and coils and circulates naturally due to convection currents inside the tank. The hot oil rises to the top of the tank and transfers heat through the walls of the tank to the surrounding air. ONAN is suitable for transformers up to about 30 MVA.

Oil natural air forced (ONAF): In this method, using fans or blowers, the air is forced to circulate around the walls of the tank and radiators. This improves heat transfer efficiency and enables higher transformer capacity (typically up to 60 MVA) compared to natural convection alone.

Oil forced air forced (OFAF): This method is used for high-capacity transformers in substations and power plants. It uses pumps to circulate oil through a heat exchanger with forced air driven by fans. This allows for the most efficient heat removal and is suitable for very large transformers. Figure 4.2 shows two examples of the location of transformer cooling devices.

4.8 POWER ELECTRONICS-BASED CONTROL

Power electronics are rapidly becoming integrated into transformer design, paving the way for a new generation of intelligent and versatile transformers known as power electronic transformers (PETs). Conventional transformers suffer from limitations such as passive operation, limited

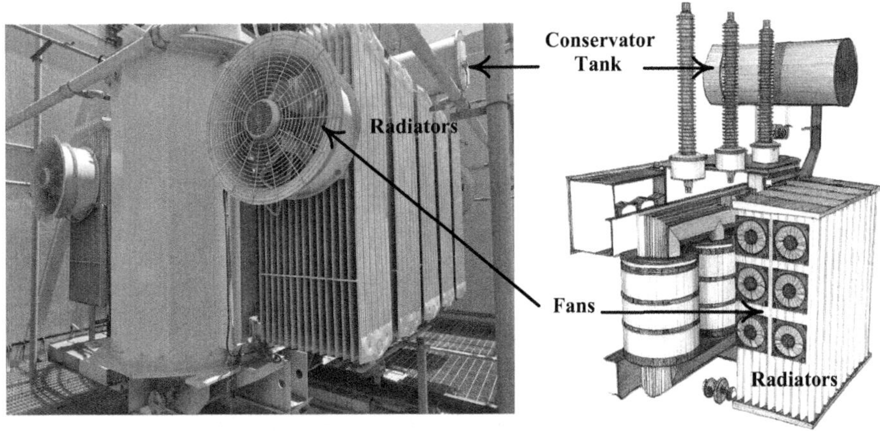

FIGURE 4.2 The location of transformer cooling devices.

operation, and large volume and weight. Traditional transformers are passive devices, meaning they can only adjust voltage and current in proportion to the number of turns in their coils. They cannot actively regulate voltage or current to a large amount. In addition, these transformers have no control over the quality of electricity or the stability of the network. Also, large iron cores and heavy copper windings have greatly increased the size and weight of conventional transformers.

Power electronics technology provides active control capabilities to transformers using power semiconductor devices such as thyristor and IGBTs. Complex algorithms that regulate power electronics technology give PET transformers the capacity to regulate voltage dynamics, enhance power quality, and boost grid stability. PETs can ensure constant voltage delivery to sensitive equipment even with variable input voltages. In addition, by filtering harmonics and reactive power control, PETs can help transfer energy cleaner and more efficiently. Power electronics enable real-time control of power flow in the grid and improve overall system stability and flexibility. Finally, using power electronics, PETs are smaller and lighter than their conventional transformer counterparts.

PETs can be divided into two categories: hybrid power electronic transformers (HPET) and modular multilevel converters (MMC). In HPETs, traditional transformers are combined with power electronics to achieve a balance between cost, efficiency, and controllability. However, MMCs are all power electronics, have high controllability due to complex designs, and are currently more expensive than HPETs.

The cost of power electronic components is very high compared to conventional transformers, and power electronics increase losses. However, continuous advances in power electronics technology make PETs a more viable option. By reducing costs and improving efficiency, we can expect a wider adoption of PETs in the future and make the power grid smarter and more dynamic.

4.8.1 Advanced Power Electronic Converters

The subject of this section is that transformers do not have the ability to actively control the current or convert between AC and DC, so advanced converters can be used. These electronic devices are more complex, flexible, and functional and are important for integrating renewable energy sources such as solar and wind into the grid. In addition, inverters can simultaneously filter harmonics and control reactive power.

Multi-level power converters, matrix converters, and DC–DC converters are among the types of advanced converters that are used with transformers or independently. Multilevel converters use

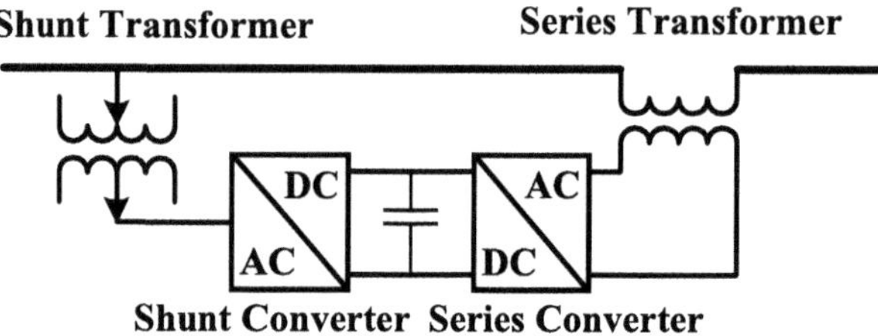

FIGURE 4.3 A UPFC as one of the FACTS devices.

multiple switching stages to produce a high-quality AC output waveform from a DC source, ideal for renewable energy grid integration. High-frequency inverters allow smaller and lighter transformer designs while maintaining efficiency. Matrix converters are complex converters that can directly convert AC to AC with variable voltage and frequency, providing more control over power flow. Finally, DC–DC converters regulate the DC voltage level, which is very important for battery storage systems and the integration of DC loads into the grid.

Finally, it should be said that while transformers still play a vital role, advanced converters play an important role in the developing power grid. With the increasing integration of renewable energies and increasing demand for flexible power conversion, we can expect further advances in these technologies and shape a more efficient, sustainable, and dynamic power system.

4.8.2 FACTS (Flexible AC Transmission System)

FACTS devices (flexible alternating current transmission systems) are independent electronic systems that interact with the transmission network and transformers to improve power flow control, stability, and overall grid efficiency. FACTS devices actively manage and optimize power transmission with inductors, capacitors, and power electronics in the transmission network and are typically located in the transmission network, often near transformers or substations.

FACTS devices interact with grid transformers in several ways. Among FACTS devices, Static VAR Compensators (SVC) can inject or absorb reactive power into the grid to help maintain voltage stability and optimize power transmission capabilities. Thyristor-controlled series compensators (TCSCs) can be placed in series with transmission lines to improve their impedance characteristics and increase power flow. Interphase power flow controllers (IPFCs) can control the power flow between specific lines within a network and improve overall network management. FACTS devices can facilitate the integration of renewable energy sources such as wind and solar manage their variable output power and ensure grid stability. The static synchronous compensator (STATCOM) can also provide more controllability with a faster response time. Finally, a unified power flow controller (UPFC) [by far the most comprehensive FACTS device] provides fast-acting reactive power compensation on HV transmission networks. It uses two transformers (one in series and the other in parallel) and a pair of three-phase controllable bridges to produce current that is injected into a transmission line using the series transformer. The controller can control active and reactive power flows in a transmission line. Figure 4.3 shows an example of a FACTS device.

In short, it can be said that FACTS devices, along with transformers play an important role in optimizing and strengthening the modern power grid.

4.9 PHASE BALANCING

Balanced operation is vital for the optimal operation of a three-phase transformer. In the state of phase balance, the voltage magnitudes are equal, their angles are ±120, and this condition is valid for currents. The symmetry of all network components causes the balance of voltages and currents, and the asymmetry of even one of the network components causes an imbalance in voltages and currents. We also know that the sum of the balanced currents becomes zero and no current passes through the neutral wire and imbalance will cause current to pass through the neutral wire. Therefore, in the case of imbalance, transformers that are not designed for such loads may be damaged. In addition, the imbalance of the currents causes the imbalance of the magnetic fluxes of the phases, which itself increases the core losses.

Single-phase loads, non-linear loads, and damaged three-phase loads are the most important factors of network imbalance, and from the point of view of the transformer, this imbalance must be dealt with. The first method is to manage and divide single-phase loads between three phases in the network so that from the point of view of the transformer, the loads are seen as three symmetrical phases. The second method is to use *phase shift transformers (PSTs).* These specialized transformers can change the phase angle of certain loads and help with the overall phase balance. The third method is to use *variable load controllers (VLCs).* These devices can automatically modify the energy consumption of individual loads to achieve balanced currents in each phase. The fourth method is to use harmonic reduction techniques. Filtering the generated harmonic distortions can have an important effect on the phase balance.

In summary, balanced phases help to improve efficiency, increase transformer life, improve network stability, and reduce maintenance costs.

Part Two: Answer-Question

4.10 TWO-CHOICE QUESTIONS (YES/NO)

1. Ideal transformers would have zero voltage regulation.
2. A higher voltage regulation percentage indicates a better ability to maintain a constant voltage.
3. Resistance, leakage reactance, and load current are effective factors in voltage regulation.
4. Off-circuit tap changers and on-load tap changers are used to improve the voltage regulation in transformers.
5. A tap changer is a mechanism outside a transformer that allows the adjustment of the turn ratio in its windings.
6. In NLTC or de-energized circuit tap changers, the transformer needs to be disconnected from the power source before adjustments can be made.
7. NLTCs are more complex and expensive and suitable for situations that require frequent voltage regulation.
8. OLTCs are simple and cheap, but they provide little control and flexibility.
9. Tap changers are found in distribution, transmission, and industrial transformers.
10. Sources of heat generation in transformers can be considered copper losses in windings, core losses, and stray losses.
11. Transformer insulation materials degrade faster at higher temperatures and this can lead to premature aging and possible failures.
12. The natural convection method is suitable for larger transformers.
13. In forced air cooling, fans are used to circulate air more effectively.
14. For smaller transformers, pumps are used to pass the oil through the radiators, improving heat transfer efficiency.
15. Temperature sensors, load management, and regular maintenance can be used to monitor and maintain the thermal performance of transformers.
16. The most important use of temperature data is to determine maintenance intervals for cleaning cooling systems and replacing old oil.

17. Thermostats manually activate cooling systems such as fans or pumps when the temperature exceeds a predetermined limit.
18. Transformer overload causes heat increase and as a result, insulation destruction, efficiency reduction, core and winding deformation, and finally shorten the life of transformers.
19. Strategies such as reducing and temporarily cutting off non-critical loads can prevent overload.
20. The first solution for overload can be upgrading the transformer to a higher capacity.
21. Remote control and monitoring increase efficiency, enable preventative maintenance, improve decision-making, reduce costs, and provide greater security.
22. The sensors collect data on temperature, oil level, pressure, vibration, and electrical characteristics and send it to the DAU.
23. DAU converts the raw data into a digital format suitable for transmission.
24. It should be said that with remote access, the importance of cyber security measures increases a lot.
25. Encryption techniques are not important to protect the system from cyberattacks.
26. The choice of cooling method is independent of the size, capacity, and application of the transformer.
27. The natural convection (NC) method is economical but limits the capacity of the transformer.
28. Oil Natural Air Natural (ONAN) is suitable for transformers up to about 30 MVA.
29. Oil Natural Air Forced (ONAF) method using fans or blowers, the air is forced to circulate around the walls of the tank and radiators.
30. The Oil Forced Air Forced (OFAF) method is used for low-capacity transformers in substations and power plants.
31. Conventional transformers suffer from limitations such as passive operation, limited operation, and large volume and weight.
32. Power electronic transformers (PETs) are a new generation of intelligent and versatile transformers.
33. Traditional transformers are passive devices, meaning they can only adjust voltage and current in proportion to the number of turns in their coils.
34. PET transformers are not able to control voltage dynamics, improve power quality, and increase grid stability.
35. Using power electronics, PETs are larger and heavier than their conventional transformer counterparts.
36. In HPETs, traditional transformers are combined with power electronics to achieve a balance between cost, efficiency, and controllability.
37. PETs face many challenges including cost and casualties.
38. Inverters can simultaneously filter harmonics and control reactive power.
39. Multilevel inverters use multiple switching stages to produce a high-quality AC output waveform from a DC source.
40. Matrix converters are complex converters that can directly convert AC to AC with variable voltage and frequency.
41. FACTS devices are often near transformers or substations.
42. Static VAR Compensators (SVC) can inject or absorb active power.
43. Thyristor-controlled series compensators (TCSCs) can increase power flow.
44. Interphase power flow controllers (IPFCs) can control the power flow between specific lines within a network and improve overall network management.
45. The static synchronous compensator (STATCOM) can also provide more controllability with a faster response time.
46. By far the most comprehensive FACTS device is UPFC.
47. In the state of phase balance, the voltage magnitudes are equal and their angles are zero.

48. The asymmetry of even one of the network components causes an imbalance in voltages and currents.
49. In the case of imbalance, transformers that are not designed for such loads may be damaged.
50. Single-phase loads, non-linear loads, and damaged three-phase loads are the most important factors of network imbalance.
51. The phase shift transformers can change the phase angle of certain loads and help with the overall phase balance.

4.11 KEY ANSWERS TO TWO-CHOICE QUESTIONS

Yes	1,3,4,6,9–11,13,15,16,18,19,21–24,27–29,31–33,36–41,43–46,48–51
No	2,5,7,8,12,14,17,20,25,26,30,34,35,42,47

5 Transformer Testing, Diagnostics, and Protection

Part One: Lesson Summary

5.1 ROUTINE TESTS

Like all crucial pieces of power system equipment, transformers should be checked regularly. Routine tests are necessary for a transformer's maintenance program and aid in revealing potential issues early enough to prevent expensive breakdowns. The purpose of these tests is to identify potential issues before they escalate into serious faults and to prevent expensive repairs, power interruptions, and possible safety threats. Additionally, frequent testing of transformer parameters allows for quicker detection of developing issues and slight changes in the key performance characteristics of the transformer – efficiency, voltage regulation, and capacity. The frequency of routine testing is determined by various factors such as age, loading conditions, and the transformer's criticality. In conclusion, the benefits of routine testing are listed in Table 5.1.

In the end, the advantages of routine testing can be summarized as follows.

- Early detection of problems: With routine testing, potential issues can be identified at an early stage; this allows for prompt repairs and minimal downtime.
- Extended transformer life: Recognizing problems early helps to make a transformer last longer by using routine tests.
- Improved grid reliability: Regular maintenance of transformers contributes to a more reliable and stable power grid.

TABLE 5.1
The Types of Routine Tests Performed on Transformers

Test	Description
Visual inspection	Inspect the transformer thoroughly, looking for leaks, cracks, overheating signs, and loose connections.
Dissolved gas analysis (DGA)	Electrical transformer oil contaminants are examined using DGA. Over time, the insulating materials in electrical equipment release gases as they degrade.
Winding resistance measurement	It is important to verify the DC resistance of each winding in this procedure. An increase in resistance may indicate loose connections, corrosion, or damage to the windings.
Turns ratio test	Checking the voltage transformation ratio accuracy is very important too.
Insulation resistance (Megger) test	This test checks the insulation resistance across the windings and the transformer tank. A drop in resistance implies possible insulation deterioration.
Leakage current test	This test checks for any current leakage that may be flowing between the windings and the tank. A high leakage current could be an indication of insulation breakdown.
Temperature monitoring	It is important to monitor the temperature of the oil and the hottest spots in the transformer to identify potential overheating problems.

DOI: 10.1201/9781003537564-5

5.2 COMMISSIONING TESTS

Perhaps you were not aware that a freshly made transformer is not immediately hooked up to the grid. Before energizing the transformer and putting it into service, commissioning tests must be carried out. With these tests, it is verified that the equipment works properly and complies with the specifications required for its use so that it can operate safely and reliably. These tests also give baseline information on certain parameters like winding resistance, insulation resistance, and oil quality. This information is important for future maintenance and monitoring. In the commissioning tests, all routine tests of Table 5.1 are performed. In addition to them, the tests in Table 5.2 are also performed.

Finally, the importance of commissioning tests can be summarized as follows:

- Safety: Commissioning tests are essential to make sure the transformer is safe and no potential dangers occur during its energization.
- Performance guarantee: These are tests that are done to show that the transformer is meeting its designed functions as well as performance specifications.
- Early problem detection: It is important to identify any problems with the transformer while carrying out its commissioning so that they can be rectified before being put into operation.

5.3 TYPE TESTS

Prior to a transformer being produced in a large or numerous amount and placed into operation, it must go through a number of type tests. These tests determine how well the transformer functions as well as its strength since it has been subjected to different operating conditions that are simulated. All these activities are geared toward ensuring that these devices can work in real-world situations by meeting industry requirements through this thorough evaluation process. Type testing is done to determine if a transformer's design complies with its stipulated voltage and current ratings, efficiency performance targets, and load ability. Such tests also aim to confirm the transformer's capability to handle mishaps like short-circuiting, overloading, and other potential faults.

Additional dielectric tests, coil resistance tests, ratio of rotation tests as well as leakage tests should be carried out on this type of testing assignment besides those stipulated in Table 5.3.

Finally, the advantages of type testing can be summarized as follows:

- Type testing helps to make the transformers more reliable and robust since it detects possible vulnerabilities during the design phase.
- It ensures that the transformer is safe in different operational situations thereby reducing chances of failure and hazards that may arise.

TABLE 5.2
The Types of Commissioning Tests Performed on Transformers

Test	Description
Polarity test	This check will verify that transformer windings have the right polarity for efficient power transfer to the grid.
Dielectric loss test (tan delta test)	During this test, a transformer's insulation dielectric loss factor is quantified to show its condition and its capability to withstand electrical stress.
Control and protection systems test	Verification of proper operation of control and protection systems, for instance, tap changers, relays, and alarms.
Cooling system functionality test	Ensuring proper operation of cooling systems like fans, pumps, and oil circulation.
Phasing checks	Confirming the correct phasing of the transformer with the existing grid.

TABLE 5.3
The Type Tests Performed on Transformers

Test	Description
Temperature rise test	Under different load conditions, this test is aimed at estimating how hot the wires and the core of the transformer become when powered. By doing this, the designers aim to guarantee that such equipment does not overheat and fail because it exceeds critical temperature levels.
Impulse voltage test	The purpose of this test is to see if lightning strikes can cause damages on an insulated transformer, which happens when the insulation is not strong enough to resist high sudden impacts from electricity. This is done to discover how well it can resist sudden high-voltage changes without being damaged.
Short circuit test	This particular test is designed to imitate a condition where there is a short circuit problem at the output terminals of the transformer whereby it measures the capacity of the transformer to survive the heavy currents found in such faults without getting damaged permanently.

- Through type testing it becomes possible to catch and rectify faults at earlier stages in order to prevent expensive redesigning after large-scale production has begun.
- Successful type testing indicates the transformer works well and is dependable which is essential in order for them to have reliance on it.

5.4 DIAGNOSTIC TESTING METHODS

Diagnostic testing methods become important for identifying potential problems early enough, avoiding expensive outages, and extending the life of a transformer. Diagnostic testing methods help in the early identification of problems that may be corrected within shorter times as part of repair and maintenance procedures for reduced downtimes and costs associated with repairs. These diagnostic tests can identify hidden problems before they escalate into major breakdowns, preventing potential safety hazards and extensive damage. In addition, operators can use diagnostic tests to understand the transformer's health. Table 5.4 encompasses various diagnostic tests that are performed to assess different transformer health parameters.

Several factors, such as transformer age and condition, suspected issues, cost, and availability determine the selection of diagnostic tests. Finally, we can say that there are many advantages to regular diagnostic testing. The most significant of them is the early identification of problems and the possibility to take corrective measures before they escalate into major issues. Diagnostic test results are also useful in making maintenance decisions and allocating resources. Moreover, the transformer can last longer if preventive maintenance is based on diagnostic test results.

5.5 OVERCURRENT PROTECTION

While transformers are meant to manage certain load capacities, unexpected events may result in too much current flowing through the transformer which may cause either harm to it or some safety risks. Safeguarding these crucial grid components is very important through overcurrent protection. This situation involves two factors, overloading and the short circuit. The transformer is overloaded if there is more current passing through it than is specified for too long in a period. This overloading may result from high load requirements or equipment malfunctions. At the forefront of the protective measures from the damaging currents in the transformers are the overcurrent protection devices. Such devices are meant to cut off "interrupt faulty currents" or "prevent overloads".

If there is a short circuit, the transformer is protected from the power source by the protection device so that the high current lasts for a very short moment and thus less damage is done. Some protective devices are designed such that they sense prolonged overloads hence allowing them to take some corrective measures including breaking circuits or reducing/eliminating loads in order to avoid overheating.

TABLE 5.4
The Diagnostic Tests Performed on Transformers

Test	Description
Dissolved gas analysis (DGA)	Following the explanation of Table 5.1, we examine oil samples from the transformer to find out if there are any traces of dissolvable gases as well as their proportions. Different types of faults such as overheating, arcing, or early stages of cellulose breakdown are indicated by particular gas compositions in it.
Furanic compound analysis	This examination looks into the existence of furanic compounds in the oil. These are referred to as decomposition products of the insulating materials in a transformer. High levels may reflect possible degradation of insulation.
Sweep frequency response analysis (SFRA)	This non-invasive test compares the transformer's frequency response against a historical baseline or a reference unit. Deviations can imply internal issues such as movement of the winding, core damage, or loose connections.
Partial discharge (PD) testing	These tests recognize and gauge partial discharges in the insulation of a transformer, which may be a warning sign to a more chronic issue. The methods for testing for partial discharge involve ultrasonic methods or electrical methods.
Oil analysis	Routine oil analysis looks at things such as consistency, moisture, and acid content. This could show that there are some problems with the oil or the internal components of the transformer in case of any changes in these parameters.
Temperature monitoring	Continuous observation of the temperatures of oil and windings helps identify overheated places, which may result in rapid aging, as well as in the formation of cracks in insulations.

Fuses, circuit breakers, and relays are types of overcurrent protection devices for transformers. When selecting an overcurrent protector, it is vital to consider how overcurrent protection devices in various sections of the power system can be coordinated so that faults can be detected at the correct locations, minimizing disruption in affected parts.

Transformers are protected against overheating, short-circuit damages, and possible fires using overcurrent. Protection devices rapidly eliminate faults hence avoiding major interruptions to the power system and enhancing stability. Figure 5.1 shows a symbolic example of overcurrent protection.

5.6 DIFFERENTIAL PROTECTION

Defects present within a transformer can result in devastation that costs much. The usage of differential protection causes for protection of transformers from internal defects that are highly sensitive and reliable. Differential protection depends on the basic theory that a transformer's input current will be the same as its output current during normal operation but when there is a problem within

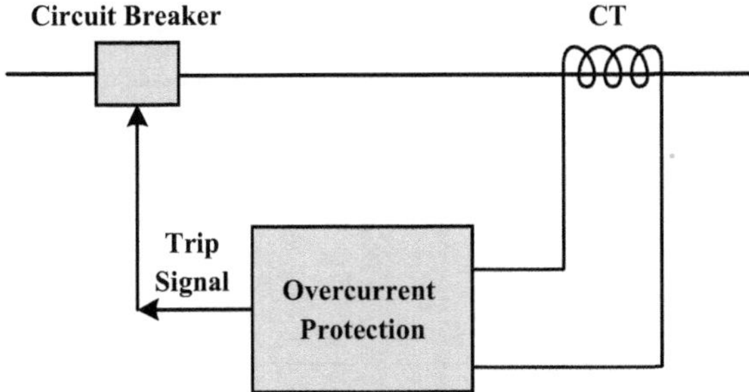

FIGURE 5.1 A symbolic example of overcurrent protection.

the unit such balance is disturbed and a difference in current results. The operation of differential protection is presented in five steps in Table 5.5.

In the end, the benefits, challenges, and considerations of differential protection are shown in Table 5.6.

Figure 5.2 shows a symbolic example of differential protection.

TABLE 5.5
How Differential Protection Works

Step	Description
1	Current transformers (CTs) are placed conveniently in both the main and secondary sides of the transformer. The high currents are stepped down to measurable levels by such CTs and the current ratio is maintained.
2	The CTs are connected in a way that creates a circulating current under normal operation. This circulating current represents the difference between the primary and secondary currents, ideally very close to zero.
3	A differential protection relay continuously monitors this circulating current.
4	When there is a fault inside the transformer the balance of current is disturbed leading to a substantial difference in circulating current that can be identified by the relay.
5	Once the fault is detected, the relay sends a trip signal which causes the primary and secondary circuit breakers of the transformer to trip-off and disconnects it from the power grid.

TABLE 5.6
Benefits, Challenges, and Considerations of Differential Protection

Benefits		Challenges and Considerations	
High sensitivity	The transformer can be safeguarded by differential protection even when small internal faults occur, it will reduce losses and downtime.	CT saturation	If there are high fault currents, the CTs can be saturated leading to wrong current measurements and malfunctioning of the relay. It is important that the selection of current transformers was done cautiously and that their sizes were correctly chosen.
Fast response	The relay operates very quickly in response to a fault, limiting the duration of the fault current and minimizing damage.	Grounding	Proper grounding of the transformer is essential for effective differential protection operation.
External fault discrimination	Differential protection does not see external faults on the power grid that would not cause a current imbalance within the transformer, preventing unnecessary tripping.	Pilot protection schemes	For higher-capacity transformers or long transmission lines, it may be appropriate to implement backup protection based on pilot communication, which helps synchronize differential relays at both ends of the transformer to improve resiliency.

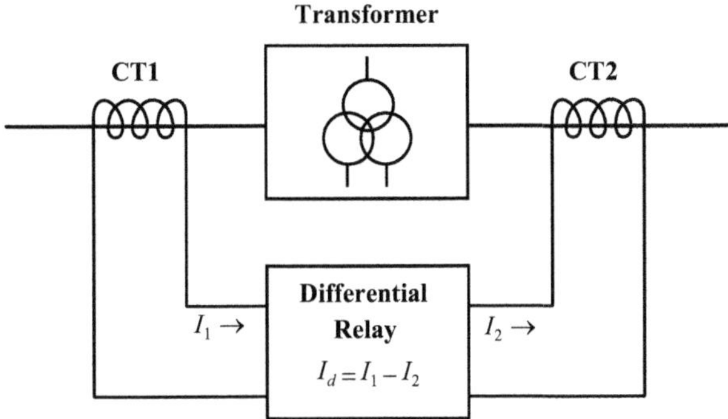

FIGURE 5.2 A symbolic example of differential protection.

5.7 SUDDEN PRESSURE RELAYS (SPRs)

Internal faults within a transformer can lead to the generation of gas due to arcing, overheating, or cellulose decomposition. Sudden pressure relays (SPRs) detect these rapid pressure surges and initiate a timely response to prevent catastrophic failures.

Short circuits or winding insulation breakdowns can generate rapid pressure increases due to arcing or rapid gas generation, and overcurrent and differential relays might not always detect these faults before they escalate, especially low-level internal faults. However, SPRs offer a complementary layer of protection by detecting rapid pressure changes, potentially indicating an internal fault before it becomes severe.

The operation steps of the sudden pressure relay are presented in Table 5.7.

SPRs are commonly used for oil-filled transformers, large power transformers, and critical transformers or transformers vital for grid stability. In the end, the benefits and limitations of SPRs are shown in Table 5.8.

Figure 5.3 shows an example of SPR in physical detail.

5.8 BUCHHOLZ RELAY

Max Buchholz invented the Buchholz relay in 1921. He was a German Luftwaffe ace and recipient of the Knight's Cross-of the Iron Cross during World War II. Buchholz relays can sense the gas presence and abnormal flow to trigger timely counteractions that prevent catastrophic failures. De-composition of insulating oil or cellulose material may generate gas within the transformer resulting from some internal faults such as short circuits, winding insulation breakdowns, or arcing. In comparison with overcurrent relays, Buchholz relays detect the presence and abnormal flow of

TABLE 5.7
How Sudden Pressure Relays Work

Step	Description
1	An SPR typically uses one or two bellows filled with gas or oil. These bellows are connected to the transformer's gas space.
2	If a fault occurs within the transformer and generates a sudden pressure rise, the bellows expand rapidly due to the increased pressure.
3	The expansion of the bellows triggers a micro-switch within the relay.
4	Depending on the configuration, the activated switch can trigger an alarm, initiate a trip signal to circuit breakers, or both. Some SPRs are set to alarm only, allowing for further investigation before a potential outage.

TABLE 5.8
Benefits and Limitations of Sudden Pressure Relays (SPRs)

Benefits		Limitations	
Early detection of internal Faults	SPRs can detect developing faults before they escalate into major breakdowns, allowing for preventive measures and minimizing damage.	False triggers	SPRs can be sensitive to external factors like earthquakes, bumps, or even other equipment malfunctions, causing unnecessary shutdowns.
Improved transformer reliability	Early fault detection and isolation help to enhance the overall reliability and lifespan of the transformer.	A trade-off between protection and security	Sometimes utilities choose to use SPRs only for alarms instead of tripping the circuit breaker entirely. This is because they can be overly cautious and cause unintended outages.
Complementary protection	SPRs work alongside other protection schemes like differential relays, providing a more comprehensive safeguard against internal faults.	Maintenance challenges	While newer SPRs are improving, older designs can be more difficult to maintain and ensure proper calibration.

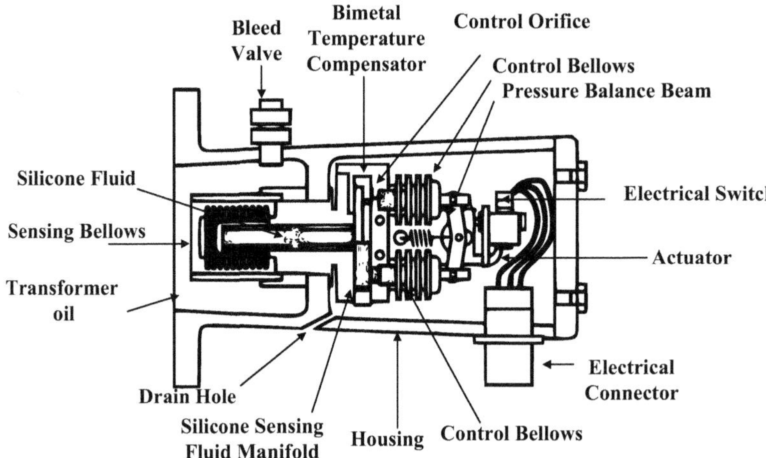

FIGURE 5.3 An example of SPR in physical detail.

gas much earlier, therefore they allow for preventive actions. The transformer is damaged extensively when faults are not detected early enough using Buchholz relays; thus both repair expenses and downtime are minimized.

Buchholz relays are commonly used in oil-filled transformers medium to large power transformers. For the operating steps in the Buchholz relay, refer to Table 5.9. Finally, the advantages and disadvantages of Buchholz relays are presented as shown in Table 5.10.

Figure 5.4 shows a symbolic figure and a real example of a Buchholz relay in detail.

5.9 SAFETY CONSIDERATIONS

Transformers contain high-voltage electricity and flammable insulating oil, necessitating a safety-first approach during installation, operation, and maintenance. Here we summarize crucial safety considerations for transformers in four sections.

TABLE 5.9
How Buchholz Relay Work

Step	Description
1	Buchholz relays are frequently placed on the piping that connects the primary transformer tank to the conservator tank, also known as an external reservoir used for storing insulating oil.
2	Any gas generated due to internal faults rises and accumulates in the conservator tank. The Buchholz relay, positioned within the connecting pipe, has a chamber that allows gas to flow through.
3	The relay contains one or two floats that are situated within the chamber. When there are no gas accumulations during normal operation, these floats stay at the top part.
4	When gas enters the chamber and accumulates, it displaces the oil, causing the floats to tilt.
5	The relay has a mercury switch that is operated by the floating elements. A mercury switch can either activate an alarm or send a signal to circuit breakers for transformer isolation depending on the extent of gas build-up and relay settings.

TABLE 5.10
Benefits and Limitations of Buchholz Relay

Benefits		Limitations	
Early warning system	Detecting gas build-up is one of the ways Buchholz relays can be used to foresee potential transformer breakdowns.	External gas intrusion	External factors like air leaks or pressure surges can introduce gas into the system, potentially leading to false alarms.
Versatility	Different types of internal faults can be detected by these relays, including small-level faults that might easily be seen by other protection schemes.	Relay placement	The performance of Buchholz relays is determined by where they are located in the piping system for correct gas collection.
Simple operation	Ones that are easy to keep in working order and very dependable are the Buchholz relay operating principles.	Limited fault diagnosis	Although gas detection is enabled by the Buchholz relay, there is still a need for further investigation due to the fact that it does not specifically identify the type of internal fault present.

5.9.1 Safety during Installation

- Qualified personnel: Only trained and qualified electrical professionals with proper safety gear should handle transformer installation.
- Site preparation: The installation site should be free from flammable materials and have adequate ventilation to prevent overheating.
- Grounding: Proper grounding of the transformer tank and neutral connections is essential to prevent stray currents and potential shock hazards.
- Lifting and placement: Use appropriate lifting equipment with the rated capacity to handle the transformer's weight safely. Secure the transformer firmly on its foundation to prevent accidental movement.

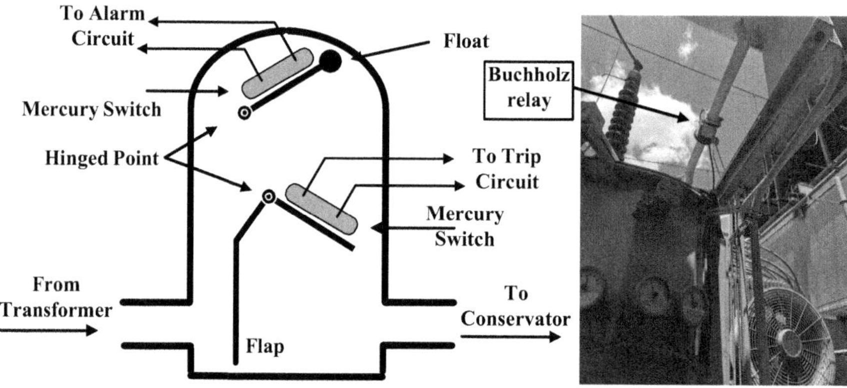

FIGURE 5.4 A Buchholz relay in detail.

5.9.2 Safety during Operation

- Warning signs: Post clear warning signs around the transformer to alert personnel of potential electrical hazards and restricted access areas.
- Security measures: Implement measures to prevent unauthorized access to the transformer, especially in outdoor locations.
- Monitoring: Regularly monitor transformer parameters like temperature, pressure, and oil level to identify any potential problems early on.
- Preventive maintenance: Schedule periodic preventive maintenance to check for leaks, loose connections, or signs of deterioration to minimize the risk of breakdowns.

5.9.3 Safety during Maintenance

- De-energization: Always ensure the transformer is completely de-energized and lockout procedures are followed before commencing any maintenance work.
- Grounding and shorting: Ground the transformer and short-circuit the high and low-voltage terminals to prevent accidental energization.
- Personal protective equipment (PPE): Wear appropriate PPE like insulated gloves, safety glasses, and flame-retardant clothing while working on transformers.
- Oil handling: Insulating oil can be flammable and pose health risks. Handle oil carefully, following proper disposal procedures for any waste oil.
- Spill response: Have a plan and designated materials readily available to address any potential oil spills.

5.9.4 Additional Safety Considerations

- Fire protection: Maintain a fire extinguisher near the transformer, suitable for electrical fires.
- Arc flash hazards: Be aware of potential arc flash hazards during maintenance and take necessary precautions like using arc flash suits and facial protection.
- Confined space entry: If transformer enclosures require entry for maintenance, follow confined space entry procedures to ensure worker safety.

By adhering to these safety considerations during installation, operation, and maintenance, utilities and personnel can ensure the safe and reliable operation of transformers, minimizing the risk of accidents and protecting human life. Remember, a commitment to safety is an investment in the well-being of workers and the continued smooth operation of the power grid.

5.10 LONGEVITY AND RELIABILITY

Understanding factors that influence transformer longevity and reliability is crucial for maximizing their lifespan and ensuring a stable power supply. The key factors affecting transformer longevity and reliability are design and manufacturing, installation and commissioning, operation and maintenance, and monitoring and diagnostics. Next, we explain each section in more detail.

5.10.1 Design and Manufacturing

- High-quality materials: Transformers built with robust materials like grain-oriented silicon steel cores and high-conductivity copper windings offer better longevity compared to those with lower-grade materials.
- Effective cooling system: A well-designed cooling system, whether air-natural, air-forced, or oil-based, helps maintain optimal operating temperatures and prevents overheating, a significant contributor to insulation degradation.
- Conservative design margins: Transformers designed with some buffer in their voltage and current ratings can handle slight overloads without excessive stress, promoting longevity.

5.10.2 Installation and Commissioning

- Proper installation: Following recommended installation practices, including proper grounding and connection tightening, ensures optimal performance and minimizes the risk of future problems.
- Thorough commissioning tests: Performing comprehensive commissioning tests verifies the transformer's functionality, identifies any potential issues early on, and establishes baseline data for future monitoring.

5.10.3 Operation and Maintenance

- Load management: Avoiding overloading transformers and operating within their designed capacity helps prevent overheating and premature aging.
- Regular maintenance: Implementing a routine maintenance program that includes inspections, oil analysis, and dissolved gas analysis (DGA) allows for early detection of potential problems and timely corrective actions.
- Environmental conditions: Minimizing exposure to excessive humidity, dust, or contaminants helps maintain insulation integrity and prevent corrosion.

5.10.4 Monitoring and Diagnostics

- Continuous monitoring: Continuously monitoring key parameters like temperature, pressure, and oil level allows for early identification of potential issues before they escalate.
- Diagnostic testing: Regular diagnostic testing with tools like DGA and SFRA (Sweep Frequency Response Analysis) provides valuable insights into the transformer's internal health and helps predict potential problems.

Table 5.11 shows the strategies for extending transformer life and the benefits of ensuring transformer longevity and reliability.

Utilities can make sure that their transformers will last for a long time and are reliable by giving the utmost importance to design quality, correct installation and operation, and maintenance plan, which is complete. Safety power supply, decreasing downtime hazards as well as extending the duration are dependent on regular monitoring, diagnosis, and database-led servicing.

TABLE 5.11
Strategies and Benefits for Extending Transformer Life

Strategies		Benefits	
Derating	Operating the transformer below its nameplate capacity can significantly extend its lifespan.	Reduced costs	Extending transformer life reduces the need for premature replacements, saving on capital expenditure.
Periodic overhaul	Planned overhauls involving cleaning, insulation testing, and replacement of aged components can rejuvenate transformers and extend their operating life.	Improved grid stability	Reliable transformers contribute to a more stable and efficient power grid, minimizing outages and disruptions.
Advanced monitoring systems	Implementing online monitoring systems with features like fault diagnostics and remote access can facilitate proactive maintenance and early problem detection.	Enhanced safety	Proper maintenance and timely replacement of aging transformers minimize the risk of catastrophic failures and potential safety hazards.

Part Two: Answer-Question

5.11 TWO-CHOICE QUESTIONS (YES/NO)

1. Routine tests are necessary for a transformer's maintenance program.
2. The main purpose of routine tests is to identify potential issues before they escalate into serious faults.
3. The frequency of routine tests is fixed for all transformers.
4. In visual inspection, we are looking for leaks, cracks, overheating signs, and loose connections.
5. Different types of internal faults within a transformer can be shown by specific gas compositions.
6. An increase in resistance may indicate loose connections, corrosion, or damage to the windings.
7. The test (Megger) is to determine the accuracy of the number of turns of the coil.
8. Recognizing problems early helps to make a transformer last longer by using routine tests.
9. A freshly made transformer is immediately hooked up to the grid.
10. In the commissioning tests, all routine tests are performed.
11. The Tan delta test will verify that transformer windings have the right polarity for efficient power transfer in the grid.
12. Type tests aim to confirm the transformer's capability to handle mishaps like short-circuiting, overloading, and other potential faults.
13. Type tests, commissioning tests, and routine tests have no tests in common.
14. The purpose of the impulse voltage test is to see if the lightning strikes can cause damage to an insulated transformer.
15. Diagnostic testing methods become important for identifying potential problems early enough.
16. High levels of furanic compounds may reflect possible degradation of insulation.
17. Sweep frequency response analysis (SFRA) is a routine test.
18. Partial discharge (PD) testing is a commissioning test.
19. Several factors, such as transformer age and condition, suspected issues, cost, and availability determine the selection of diagnostic tests.

20. Overcurrent protection involves two factors, overloading and the short circuit.
21. Differential protection is meant to cut off "interrupt faulty currents" or "prevent overloads".
22. Overcurrent protection depends on the basic theory that a transformer's input current will be the same as its output current during normal operation.
23. The CTs of the transformer in the differential relay are connected in a way that creates a circulating current under normal operation.
24. Sudden pressure relays (SPRs) detect gas rapid pressure surges.
25. Overcurrent and differential relays always detect short circuits or winding insulation breakdowns before they escalate, especially low-level internal faults.
26. Buchholz relays can sense gas presence and abnormal oil flow.
27. Buchholz relays are placed on the piping that connects the primary transformer tank to the conservator tank.
28. The Buchholz relay has a mercury switch that is operated by the floating elements.
29. Transformers designed with some buffer in their voltage and current ratings can handle slight overloads without excessive stress, promoting longevity.

5.12 KEY ANSWERS TO TWO-CHOICE QUESTIONS

Yes	1,2,4–6,8,10,12,14–16,19,20,23,24,26–29
No	3,7,9,11,13,17,18,21,22,25

6 Emerging Technologies in Transformers for Smart Grid Integration

Part One: Lesson Summary

6.1 DIGITAL TRANSFORMERS

Conventional transformers are reliable but lack the intelligence needed for a modern and dynamic grid. Digital transformers bridge this gap by incorporating sensors, communication modules, and intelligent controllers. Conventional transformers are being infused with intelligence to create a new breed, which is the so-called "digital transformers".

Sensors gather real-time data on various parameters like temperature, voltage, current, and vibration. Communication modules enable transformers to communicate with central control systems, sharing valuable data insights, and intelligent controllers process sensor data and make informed decisions to optimize performance and predict potential issues.

The advantages of digital transformers can be summarized as follows:

- Enhanced efficiency: Real-time monitoring and data analysis allow for optimal operation, minimizing energy losses and maximizing efficiency.
- Improved reliability: Predictive maintenance capabilities help identify and address potential problems before they occur, reducing downtime and improving grid reliability.
- Smarter grid integration: Digital transformers can integrate seamlessly with smart grid technologies, enabling features like dynamic voltage control and improved grid stability [43–45].
- Renewable energy integration: Their capability to manage the variable power harvested from renewable sources like solar and wind enables a smoother integration into the grid.
- Data-driven decision-making: Digital transformers provide valuable data insights that can inform grid management decisions and optimize asset utilization.

Digital transformers do not operate in isolation. They function within a broader digital ecosystem that includes smart grid communication networks, data analytics platforms, and cybersecurity measures. Smart grid communication networks enable secure and reliable data exchange between transformers, control centers, and other grid devices. In addition, data analytics platforms process and analyze transformer data, providing actionable insights for grid operators. Finally, robust cyber security measures must be taken into account to protect the digital ecosystem against cyber-attacks that can disrupt grid operations.

While digital transformers offer significant advantages, there are challenges to address. In addition, despite the challenges, the adoption of digital transformers is accelerating. As technology advances, costs decrease, and cybersecurity measures improve, digital transformers are ready to revolutionize the power grid. Table 6.1 shows the advantages and challenges of digital transformers.

Digital transformers represent a paradigm shift in power grid technology. Their ability to enhance efficiency, improve reliability, and integrate renewable energy sources paves the way for a smarter,

DOI: 10.1201/9781003537564-6

TABLE 6.1
The Advantages and Challenges Associated with Digital Transformers

Advantages		Challenges	
Self-healing grids	Digital transformers could become intelligent nodes within the grid, autonomously adjusting their operation and identifying maintenance needs, leading to self-healing grids with minimized downtime.	Cost	The initial investment in digital transformers and the supporting infrastructure can be higher compared to traditional options.
Real-time optimization	Advanced data analysis from digital transformers could optimize power flow and distribution in real time, ensuring efficient and reliable electricity delivery.	Data security	Securing the vast amount of data generated by digital transformers is critical to prevent cyberattacks.
Sustainability and resilience	By integrating renewable energy sources and enabling efficient grid management, digital transformers can contribute to a more sustainable and resilient power grid for the future.	Interoperability	Standardization of communication protocols and data formats is essential for seamless integration with existing grid infrastructure.

more sustainable, and more resilient power grid of the future. As this technology matures and overcomes existing obstacles, digital transformers are set to become the main players in a transformed power grid landscape.

6.2 INTEGRATION WITH SMART GRIDS

The modern power grid is undergoing a significant evolution. Smart grids, characterized by two-way communication, automation, and real-time monitoring, are being implemented to address challenges like growing energy demands, integration of renewable energy sources, and increasing concerns about efficiency and reliability. In this evolving landscape, transformers play a crucial role, and their integration with smart grids unlocks a range of benefits [43–45].

Conventional transformers inherently lack the intelligence needed for a smart grid. In contrast, smart transformers represent a new generation, equipped with sensors, communication modules, and intelligent controllers. In short, the advantages of a smart transformer and the advantages of integrating transformers with smart grids can be summarized in Table 6.2.

While the advantages are compelling, there are several challenges to address. Standardization of communication protocols and data formats is crucial for seamless integration of transformers from different manufacturers. In addition, the increased reliance on communication technologies necessitates robust cybersecurity measures to protect against cyberattacks. In the end, even though, the initial investment in smart transformers and the communication infrastructure can be significant, despite the challenges, the integration of transformers with smart grids is well underway. As technology advances, costs decrease, and cybersecurity measures improve, smart transformers will play an increasingly vital role in building a more efficient, reliable, and sustainable power grid of the future [46–54].

6.3 HIGH-TEMPERATURE SUPERCONDUCTING (HTS) TRANSFORMERS

Imagine a transformer that boasts significantly reduced energy loss, a compact footprint, and environment-friendly operation. This is the promise of High-Temperature Superconducting (HTS) transformers, a revolutionary technology dignified to transform the power grid landscape. HTS transformers utilize special high-temperature superconducting materials that lose all electrical

TABLE 6.2
The Key Advantages of a Smart Transformer

Monitor key parameters	Real-time data on voltage, current, temperature, and other critical metrics is collected.
Optimize power flow	The collected data allows for dynamic adjustments to voltage levels, minimizing energy losses and improving grid stability.
Identify/prevent failures	Early detection of anomalies and potential equipment failures allows for preventive maintenance and minimizes downtime.
Facilitate integration of renewables	Smart transformers can handle the variable power output of renewable sources such as solar and wind, ensuring grid stability.
Improve situational awareness	Utilities gain a comprehensive view of the grid's health, enabling informed decision-making and proactive response to issues.
Enhanced efficiency	Smart transformers optimize power flow, reducing energy losses and minimizing overall energy consumption.
Improved reliability	Real-time monitoring and fault detection capabilities enable proactive maintenance and prevent cascading outages.
Increased grid capacity	Smart transformers can handle fluctuating power demands and integrate renewable energy sources, maximizing grid capacity.
Cost savings	Reduced energy losses, improved maintenance practices, and fewer outages translate to significant cost savings for utilities and consumers.
Sustainable power grid	Smart grid integration with transformers facilitates the use of renewable energy sources, promoting a more sustainable energy future.
Distribution automation systems	DAS enables two-way communication between transformers and the central control system.
Wireless sensor networks	WSNs Facilitate data collection from multiple sensors within the transformer.
Cybersecurity measures	Secure communication protocols are essential to protect against cyberattacks on the smart grid infrastructure.

resistance when chilled to extremely low temperatures. This allows for near-perfect current transmission, significantly reducing energy losses. To make things easier to understand, materials that reach liquid nitrogen temperature and become superconducting are referred to as high-temperature superconductors (HTS), while materials that need liquid helium cooling are referred to as low-temperature superconductors (LTS). In short, the benefits, applications, challenges, and considerations of HTS transformers can be summarized in Table 6.3.

HTS transformers represent a significant leap forward in transformer technology. While challenges remain, their potential benefits for efficiency, environmental impact, and power grid capacity are undeniable. As technology matures and costs become more competitive, HTS transformers have the potential to revolutionize the way we deliver and manage electricity in the future. Figure 6.1 shows a symbolic figure of a HTS transformer.

6.4 AMORPHOUS METAL CORES

Most transformers rely on cores made from silicon steel. While effective, silicon steel cores have limitations on energy and magnetization losses. Amorphous metals are a class of metallic materials with a unique atomic structure. Unlike the crystalline structure of silicon steel, amorphous metals are essentially solidified liquids, resulting in several advantages for transformer cores. In short, the benefits and applications of amorphous metal core transformers can be summarized in Table 6.4.

Amorphous metal cores, while promising, have disadvantages such as higher cost and limited production capacity. However, despite the challenges, advancements in manufacturing processes are continuously driving down the cost of amorphous metals. As production capacity scales up, amorphous metal cores are poised to become a more mainstream choice for transformers.

TABLE 6.3
The Benefits, Applications, Challenges, and Considerations of HTS Transformers

	Benefits
Unparalleled efficiency	HTS transformers can achieve efficiency levels exceeding 99%, compared to around 95% for conventional transformers. This translates to substantial energy savings and reduced operating costs.
Compact design	Because HTS windings experience no resistance, they can be smaller and lighter compared to traditional copper windings. This allows for more compact transformers, ideal for space-constrained urban environments.
Environmentally friendly	HTS transformers eliminate the need for oil-based cooling systems, which can pose environmental risks in case of leaks. Additionally, their high efficiency reduces overall power plant emissions.
Increased power capacity	HTS transformers can handle larger loads without overheating, enabling them to support growing energy demands in densely populated areas.
	Applications
Urban substations	Their compact size and high efficiency make them ideal for congested urban areas.
Transmission lines	HTS transformers can minimize energy loss over long-distance transmission lines.
Renewable energy integration	Their ability to handle fluctuating power output makes them suitable for integrating renewable energy sources such as wind and solar.
	Challenges and Considerations
High manufacturing cost	Currently, HTS materials are expensive to produce, making these transformers a more costly option than conventional ones.
Cooling requirements	HTS materials require cooling to very low temperatures (around – 196°C), which presents a technological challenge and adds to operational complexity and cost.
Grid integration	The integration of HTS transformers into existing power grids requires careful planning and infrastructure upgrades.

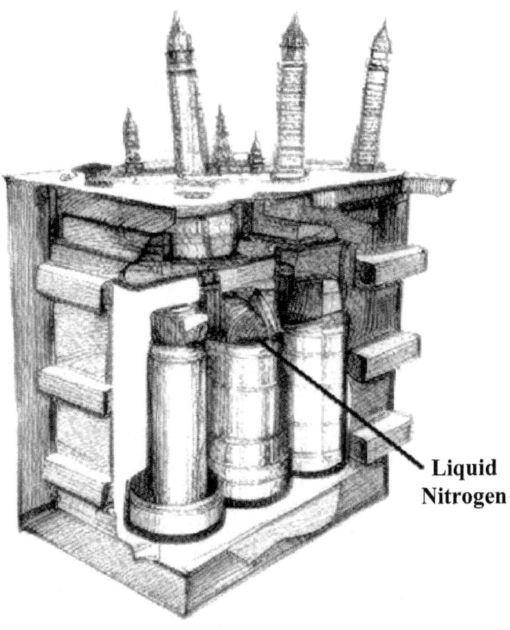

FIGURE 6.1 A symbolic figure of a HTS transformer.

TABLE 6.4
The Benefits and Applications of Amorphous Metal Core Transformers

Reduced eddy current losses	The disordered atomic structure of amorphous metals significantly reduces eddy currents, a major contributor to energy loss in traditional cores.
Lower hysteresis losses	Amorphous metals exhibit lower hysteresis losses, the energy wasted during the magnetization and demagnetization cycles.
Improved efficiency	Combined, these factors lead to transformers with significantly lower overall energy losses, translating to substantial efficiency gains.
Energy savings	Transformers with amorphous metal cores can achieve efficiency levels exceeding 99% compared to around 95% for conventional transformers. This translates to significant energy cost savings over the lifespan of the transformer.
Reduced operating costs	Lower energy losses mean less heat generation, reducing the need for cooling systems and associated maintenance costs.
Environmentally friendly	By minimizing energy consumption, amorphous metal cores contribute to a more sustainable power grid with lower greenhouse gas emissions.
Compact design	The inherent efficiency of amorphous metal cores allows for smaller transformer designs, which can be beneficial in space-constrained environments.

6.5 ADVANCED INSULATING FLUIDS

A crucial component within transformers is the insulating fluid, which performs several vital functions. Insulating fluid prevents current from flowing unintentionally between the transformer's windings, which could cause a short circuit. Insulating fluid also helps dissipate heat generated within the transformer, preventing overheating and potential damage. Finally, in case of an electrical discharge within the transformer, the insulating fluid helps extinguish the arc, minimizing damages. Traditionally, mineral oils have been the go-to insulating fluid for transformers. However, there is a growing need for alternatives due to factors like environmental concerns, fire safety, and aging and breakdown.

Therefore, researchers are exploring and developing advanced insulating fluids to address the limitations of mineral oil. Here are some promising contenders: natural esters, synthetic esters, silicone fluids, perfluorocarbons (PFC) and hydrofluorocarbons (HFC). Among these insulators, choosing the most suitable insulating fluid depends on various factors. High-capacity transformers may require fluids with superior fire resistance, while environmentally sensitive locations may favor biodegradable options. In addition, advanced fluids are more expensive than traditional mineral oils, which should be considered.

6.6 SOLID-STATE TRANSFORMERS (SSTs)

Conventional transformers rely on bulky iron cores to create a magnetic field for the sake of voltage conversion. Solid-state transformers (SSTs) take a radical leap forward, replacing these cores with sophisticated power electronics. SSTs are a transformative technology poised to reshape the power grid landscape. This shift unlocks a range of potential benefits. In short, the benefits, applications, challenges, and considerations of SSTs can be summarized as listed in Table 6.5.

Solid-state transformers represent a paradigm shift in power transmission and distribution technology. Their potential for increased efficiency, flexibility, and grid intelligence paves the way for a more sustainable and reliable power grid of the future. As SST technology matures and overcomes existing challenges, we can expect to see it transforming the way we deliver and manage electricity for generations to come. Figure 6.2 shows four symbolic examples of SST.

TABLE 6.5
The Benefits, Applications, Challenges, and Considerations of SSTs

	Benefits
Increased efficiency	SSTs can achieve efficiency levels exceeding 99%, compared to around 95% for conventional transformers. This translates to significant energy savings and reduced operating costs.
Compact design	Without bulky iron cores, SSTs can be significantly smaller and lighter than their conventional counterparts. This makes them ideal for space-constrained urban environments.
Enhanced flexibility	SSTs offer greater control over voltage and power flow, enabling them to adapt to dynamic grid conditions and integrate renewable energy sources more effectively.
Improved reliability	Solid-state electronics offer faster response times and better fault tolerance compared to traditional transformers, potentially improving grid reliability.
Bidirectional power flow	Unlike conventional transformers, SSTs can facilitate bidirectional power flow, enabling features like distributed generation and energy storage integration.
	Challenges and Considerations
Cost	The current cost of SSTs is significantly higher than conventional transformers. As the technology matures, economies of scale are expected to bring down prices.
Complexity	Solid-state electronics introduce new complexities compared to simpler transformer designs. Robust control systems and protection measures are crucial.
Grid integration	Integrating SSTs into existing power grids requires careful planning and potential upgrades to accommodate their unique characteristics.
	The Future of Solid-State Transformers
Smarter grids	SSTs can act as intelligent nodes within the grid, facilitating real-time monitoring, control, and optimization of power flow.
Renewable energy integration	Their ability to handle variable power output and bidirectional flow makes SSTs ideal for integrating renewable energy sources like solar and wind.
Improved power quality	SSTs can provide better voltage regulation and power quality, enhancing the reliability of power delivered to consumers.

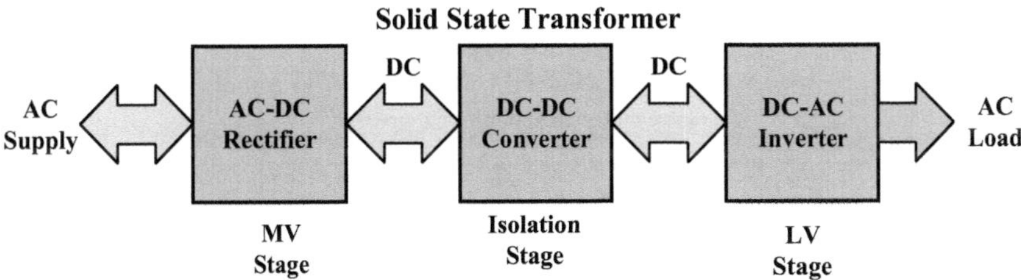

FIGURE 6.2 Four symbolic examples of SST.

6.7 MODULAR TRANSFORMERS

Modular transformers have a flexible approach to power delivery, are more adaptable, are easier to maintain, and offer a wider range of applications. Conventional transformers are reliable but they have limitations such as fixed design, scalability challenges, and maintenance complexity. Modular transformers have a building block approach.

Instead of a single, monolithic unit, Modular Transformers are comprised of standardized building blocks, power, control, and auxiliary modules. Power Modules handle the core function of voltage conversion and can be interconnected to achieve the desired voltage level and power rating. Control modules manage and monitor the operation of the power modules, ensuring efficient and safe power conversion. Auxiliary modules can provide additional functionalities like protection, metering, and communication capabilities.

In short, the benefits, applications, challenges, and considerations of SSTs can be summarized in Table 6.6.

Figure 6.3 shows a typical example of a modular transformer.

TABLE 6.6
The Benefits, Applications, Challenges, and Considerations of Modular Transformers

	The Advantages of Modularity
Flexibility and scalability	Modular transformers offer the ability to easily adjust power capacity by adding or removing power modules. This is ideal for applications with fluctuating power demands.
Simplified maintenance	Since the system is modular, faulty components can be easily identified and replaced with minimal downtime. This reduces maintenance costs and improves system availability.
Improved efficiency	Modular design allows for optimized power conversion within each module, potentially leading to higher overall efficiency
Reduced footprint	Modular transformers can be configured to fit space-constrained environments more efficiently compared to traditional designs.
	Applications of Modular Transformers
Renewable energy integration	The ability to scale power capacity dynamically makes modular transformers suitable for integrating renewable energy sources with variable power output.
Data centers	Data centers have fluctuating power demands, and modular transformers can adapt to these changes efficiently.
Microgrids	Modular transformers can be a key component in building flexible and scalable microgrids for localized power generation and distribution.
Electric vehicle charging stations	The ability to scale power capacity makes modular transformers ideal for supporting the growing demand for EV charging infrastructure.
	Challenges and Considerations
Standardization	For widespread adoption, standardized modular components, and communication protocols are crucial to ensure compatibility between different manufacturers.
Control complexity	Managing and optimizing the operation of multiple interconnected modules requires sophisticated control systems.
Initial cost	While offering long-term benefits, the upfront cost of modular transformers is generally higher than traditional options.

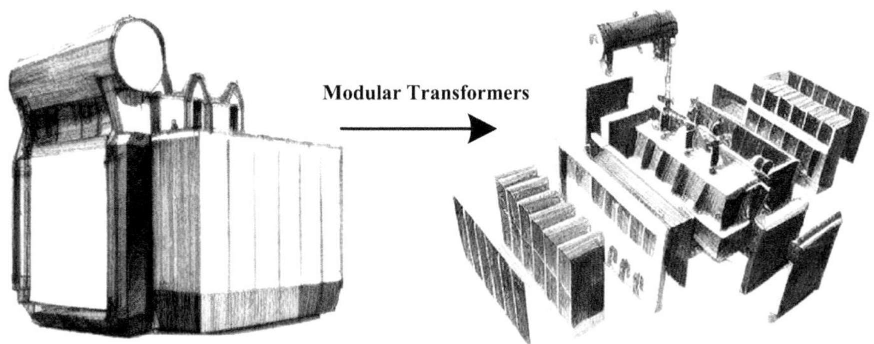

FIGURE 6.3 A typical example of a modular transformer.

6.8 DIGITAL TWINS AND ADVANCED MODELING

In the ever-evolving world of power grids, transformers play a critical role, silently stepping up and down voltage levels to ensure efficient electricity delivery. However, what if we could create digital replicas of these transformers, allowing for better monitoring, prediction, and optimization, this is where transformer digital twins and advanced modeling come into play.

Traditionally, transformer health is monitored using physical sensors that collect data on parameters such as temperature, pressure, and vibration. While valuable, this approach has limitations such as limited data points, reactive approach, and lack of predictive power. Traditional monitoring often identifies issues only after they occur, limiting opportunities for preventative maintenance and without a comprehensive understanding of a transformer's behavior, predicting future problems becomes challenging.

In other words, a transformer digital twin is "a virtual representation" of a physical transformer. It is created by combining real-world data with sophisticated modeling techniques. This digital replica can be used to simulate various operating scenarios, analyzing how the transformer would respond to changing loads, temperatures, or environmental conditions. In addition, by analyzing data and simulations, the digital twin can identify potential problems before they occur, enabling preventative maintenance and reducing downtime. As well as, the digital twin can be used to optimize transformer operation, minimizing energy losses and maximizing efficiency. In short, advanced modeling techniques, benefits, challenges, and considerations of transformer digital twins can be summarized in Table 6.7.

TABLE 6.7
The Advanced Modeling Techniques, Benefits, Challenges, and Considerations of Transformer Digital Twins

	Advanced Modeling Techniques
Finite element analysis (FEA)	The FEA technique creates a detailed 3D model of the transformer, allowing for the simulation of stress, vibration, and heat distribution within the transformer.
Computational fluid dynamics (CFD)	CFD models simulate the flow of cooling fluids within the transformer, providing insights into cooling efficiency and potential overheating risks.
Machine learning(ML)	ML algorithms can be trained on historical data and sensor readings to identify patterns and predict future transformer behavior.
	Benefits of Transformer Digital Twins
Improved reliability	Predictive maintenance and early detection of potential issues contribute to increased transformer reliability and reduced downtime.
Enhanced efficiency	Digital twins enable optimization of operating parameters, leading to improved energy efficiency and lower operating costs.
Reduced risk	By proactively identifying and addressing potential problems, digital twins help mitigate risks associated with transformer failures.
Better decision-making	Insights from the digital twin empower utilities to make data-driven decisions regarding maintenance, upgrades, and grid management strategies.
	Challenges and Considerations
Data quality	The accuracy of the digital twin depends significantly on the quality and completeness of the real-world data used to build and update the model.
Cybersecurity	As digital twins become more integrated with power grid operations, robust cybersecurity measures are essential to protect against cyberattacks.
Complexity	Developing and maintaining accurate digital twins requires expertise in modeling techniques and data analysis.

6.9 ADVANCED DATA ANALYTICS AND AI

It is clear that in the vast amount of data generated by transformers, lies a treasure trove of insights, waiting to be unlocked through the power of advanced data analytics and artificial intelligence (AI).

Modern transformers are equipped with various sensors that collect a wealth of data, including temperature, vibration, and load data. Traditional data analysis methods, while valuable, often may only analyze a single data point at a time, failing to capture the complex relationships between different parameters. Also, without advanced tools, predicting future problems can be challenging.

Advanced data analysis and artificial intelligence techniques can be considered with three methods machine learning (ML), deep learning (DL), and big data analytics (BDA).

Machine learning algorithms can learn from historical data to identify patterns and anomalies, predict potential equipment failures, and optimize maintenance schedules. Deep Learning is a specific type of ML. Deep learning can simultaneously analyze vast amounts of data from multiple sensors, uncovering hidden correlations and insights. Finally, techniques for handling large and complex datasets are crucial for processing the continuous data stream generated by transformers. In short, the benefits, challenges, and considerations of AI-powered transformer data analysis can be summarized in Table 6.8.

6.10 ADVANCEMENTS IN TRANSFORMER TECHNOLOGY

Transformer technology is not static; it is constantly evolving to meet the demands of a changing world. Table 6.9 is a glimpse of some of the exciting advancements in transformers.

Research and development in areas such as smart grid integration and environmentally friendly materials will continue to push the boundaries of transformer technology.

TABLE 6.8
The Benefits, Challenges, and Considerations of AI-Powered Transformer Data Analysis

	Benefits
Predictive maintenance	By identifying potential issues before they occur, AI can significantly reduce downtime and maintenance costs.
Improved efficiency	Insights from data analysis can help optimize transformer operation, minimizing energy losses and maximizing efficiency.
Enhanced grid management	Real-time analysis of transformer data allows for better grid management decisions, ensuring reliable and efficient power delivery.
Integration of renewables	AI can help integrate renewable energy sources by predicting power output fluctuations and optimizing grid operations accordingly.
	Challenges and Considerations
Data quality and security	The quality and security of the data collected from transformers are paramount for obtaining reliable AI models and protecting critical infrastructure.
Model interpretability	Understanding the reasoning behind AI predictions is crucial for building trust and ensuring transparency in decision-making.
Integration and expertise	Integrating AI tools with existing grid management systems and building expertise in data science and AI are essential for successful implementation.
Self-healing grids	AI-powered transformers could autonomously adjust their operation and predict maintenance needs, leading to self-healing grids that minimize downtime and improve overall reliability.
Real-time optimization	Advanced AI could analyze grid data in real time to optimize power flow and distribution, ensuring efficient and reliable electricity delivery.
Improved asset management	AI-powered insights can extend the lifespan of transformers by enabling proactive maintenance and optimized operation.

TABLE 6.9
The Exciting Advancements in Transformers

Amorphous metals	These new core materials offer lower energy losses, translating to increased efficiency and reduced operating costs.
Nano-crystalline materials	These advanced alloys promise even lower losses and improved thermal stability compared to the traditional silicon steel cores.
Biodegradable insulating oils	Environmentally conscious alternatives to mineral oil are being developed for the sake of sustainability enhancement.
High-efficiency windings	New winding techniques and materials minimize energy loss within the transformer itself.
Smart transformer monitoring	Real-time monitoring systems allow for early detection of potential issues and preventative maintenance, optimizing performance.
Digital twins	Creating digital simulations of transformers for design and testing, enabling performance prediction, operational management, and optimization strategies.
Solid-state transformers	These modern transformers use semiconductor-based power electronics instead of cores and windings, offering the potential for increased efficiency, flexibility, and fault tolerance.
On-load tap changers	These advanced tap changers can adjust voltage levels without interrupting power flow, crucial for integrating renewable energy sources with variable output.
Dry-type transformers	These transformers use air or an inert gas for cooling, eliminating the risk of oil leaks and environmental contamination.
Natural ester coolants	These biodegradable alternatives to mineral oil offer a safer and more environmentally friendly cooling solution.
Noise reduction technologies	New transformer designs and materials are being developed to minimize noise pollution in urban areas.

Advancements in this field also benefit various industries that rely on efficient power conversion, such as data centers, railway systems, electric vehicle charging stations, and renewable energy integration.

Part Two: Answer-Question

6.11 TWO-CHOICE QUESTIONS (YES/NO)

1. Conventional transformers lack the intelligence needed for a modern, dynamic grid.
2. Digital transformers are made from digital devices.
3. Digital transformers can integrate seamlessly with smart grid technologies.
4. Digital transformers operate in isolation.
5. Securing the vast amount of data generated by digital transformers is critical to prevent cyberattacks.
6. Conventional transformers play an important role in a smart grid.
7. The initial investment in smart transformers and the communication infrastructure can be significant.
8. Smart transformers can handle the variable power output of renewable sources like solar and wind, ensuring grid stability.
9. Smart grid integration with transformers facilitates the use of renewable energy sources.
10. DAS enables two-way communication between transformers and the central control system.
11. Smart transformers do not affect the optimal power flow.

12. Smart transformers have no impact on fluctuating power needs.
13. As technology matures, costs become more competitive.
14. HTS transformers can achieve efficiency levels exceeding 70%, compared to around 60% for conventional transformers.
15. HTS windings can be smaller and lighter compared to traditional copper windings.
16. Compact size and high efficiency make HTS transformers ideal for congested urban areas.
17. Currently, HTS transformers are a cheaper option compared to conventional transformers.
18. The disordered atomic structure of amorphous metals significantly reduces eddy currents.
19. Amorphous metals exhibit lower hysteresis losses.
20. Transformers with amorphous metal cores can achieve efficiency levels exceeding 70% compared to around 55% for conventional transformers.
21. Amorphous metal cores allow the design of larger transformers.
22. Perfluorocarbons (PFC) and hydrofluorocarbons (HFC) are an advanced type of insulation in transformers.
23. SSTs replacing transformer cores with sophisticated power electronics.
24. SSTs can achieve efficiency levels exceeding 99%, compared to around 95% for conventional transformers.
25. SSTs offer greater control over voltage and power flow.
26. The current cost of SSTs is significantly lower than conventional transformers.
27. Modular transformers have a building block approach.
28. In modular transformers, faulty components can be easily identified and replaced with minimal downtime.
29. In modular transformers potentially leading to lower overall efficiency.
30. The upfront cost of modular transformers can be higher than traditional options.
31. Transformers digital twins means digital copies of Transformers.
32. A transformer digital twin is a virtual representation of a physical transformer.
33. The transformer's digital twin can be used to optimize transformer operation, minimizing energy losses and maximizing efficiency.
34. Artificial intelligence techniques can be considered with three methods machine learning (ML), deep learning, and big data analytics.
35. AI can help integrate renewable energy sources by predicting power output fluctuations and optimizing grid operations accordingly.
36. AI-powered insights can extend the lifespan of transformers.

6.12 KEY ANSWERS TO TWO-CHOICE QUESTIONS

Yes	1,3,5,7–10,13,15,16,18,19,22–25,27,28,30–36
No	2,4,6,11,12,14,17,20,21,26,29

Appendix
Transformer Nameplate

Transformer nameplates provide information about the transformer's performance, specifications, key technical parameters, operating environment, and other important details. Typically, it is made of aluminum, copper, or stainless steel. These specifications are in Figures A.1–A.5. In short, these specifications can be defined as follows.

1. Rated power: The rated power is specified in kilovolt-amperes (kVA). This indicates the power capacity of the transformer, e.g., 50 kVA.
2. Primary and secondary voltage: The input and output voltages of the transformer. For example, a primary voltage of 20 kV and a secondary voltage of 400 or 220 V.
3. Rated current: The rated current on the primary and secondary sides, specified in amperes (A). The rated current is usually calculated based on the rated power and voltage.
4. Winding configuration: The type of winding connections for the primary and secondary sides, such as star (Y) or delta (Δ). For example, Y/Δ.
5. Frequency: The operating frequency of the transformer, specified in hertz (Hz). Commonly, this is 50 or 60 Hz.
6. Insulation class: The type and grade of insulation used in the windings, usually determined by standards like IEC or ANSI.
7. Temperature rise: The allowable temperature rise of the windings and transformer oil when operating under rated conditions.
8. Impedance: The impedance of the transformer, usually expressed as a percentage of the rated voltage.
9. Turns ratio: The ratio of the number of turns in the primary winding to the number of turns in the secondary winding, determining the voltage ratio between the primary and secondary sides.
10. Losses: The load and no-load losses, which correspond to losses due to the load current and core losses, respectively.
11. Cooling type: The type of cooling system used, such as air-cooled (AN) or oil-cooled (ONAN).
12. Weight: The weight of the transformer, including the core, windings, oil, and enclosure.
13. Dimensions: The physical dimensions of the transformer, including length, width, and height.
14. Noise level: The level of noise produced by the transformer, measured in decibels (dB).

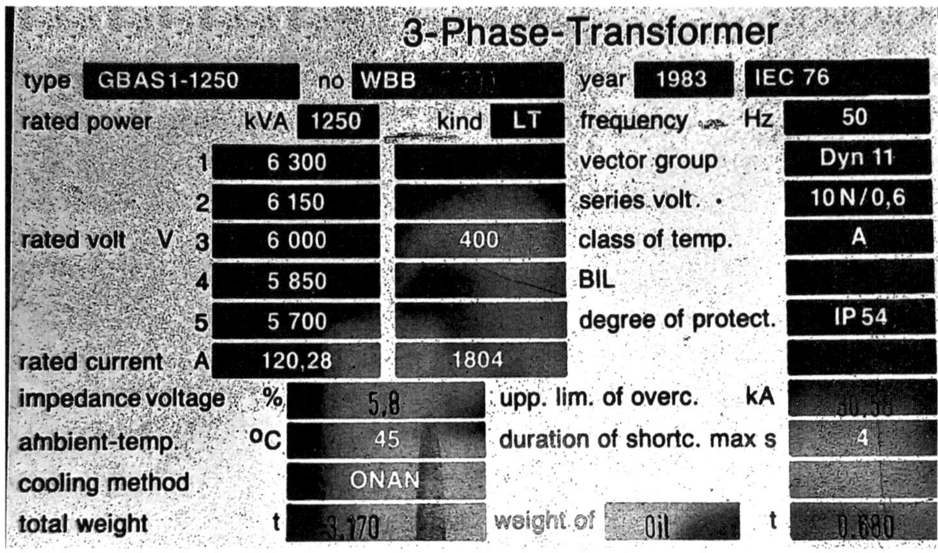

FIGURE A.1 The specification nameplate of a 1,250 MVA transformer.

3-Phase Generator Transformer 4AT 01

Type TSF 192600 Serial number GM 101 922 Number of the specification IEC 76/1-5

Year of manufacture 1984 Connection symbol ⅄ / ◁ Insulation level 1)

	System 1	System 2
Rated power kVA	176000	176000

Rated frequency Hz 50

Position	Rated voltage System 1	Rated voltage System 2	Rated current System 1	Rated current System 2	Impedance voltage ref. to rated power system 1 System 1/2	
1	V 161450		A 629.4		% 13.7	
10	V 143000	11500	A 710.6	8836	% 12.7	
26	V 110200		A 922.0		% 11.1	

	System 1	System 2
Sym. short circuit current kA	6.5	80.4
Short circuit duration max. s	2	2

Type of cooling ODAF Temp. rise oil/winding K 56.6 61.6 Insulating liquid Shell Diala D

Approx. weights:

Total weight t 140 Transportation weight t 129 Weight of oil t 32.5

Untanking weight t 82

On-load tap changer CRNG 1403/1232 Insulation voltage class kV 140 Rated current A 1200

1) LI 550 AC 230-LI 95 AC 38/LI 95 AC 38

FIGURE A.2 The specification nameplate of a 176 MVA transformer.

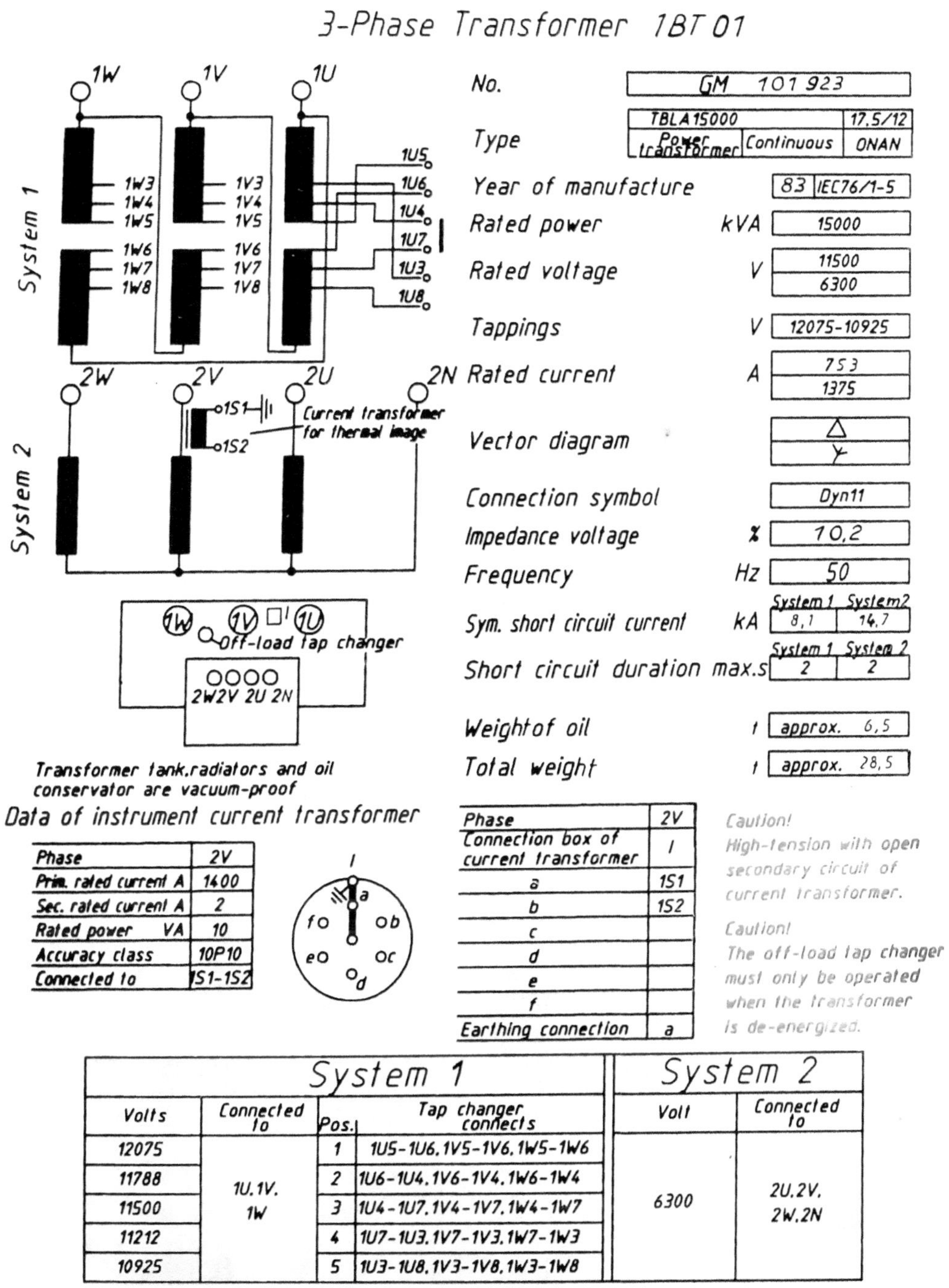

3-Phase Transformer 1BT 01

1W 1V 1U
1W3 1W4 1W5 1W6 1W7 1W8
1V3 1V4 1V5 1V6 1V7 1V8
1U5 1U6 1U4 1U7 1U3 1U8
System 1
2W 2V 2U 2N
1S1 1S2
Current transformer for thermal image
System 2
1W 1V 1U
Off-load tap changer
2W 2V 2U 2N

Transformer tank, radiators and oil conservator are vacuum-proof

No.		GM 101 923
Type		TBLA 15000 17,5/12 Power transformer Continuous ONAN
Year of manufacture		83 IEC76/1-5
Rated power	kVA	15000
Rated voltage	V	11500 / 6300
Tappings	V	12075-10925
Rated current	A	753 / 1375
Vector diagram		Δ / Y
Connection symbol		Dyn11
Impedance voltage	%	10,2
Frequency	Hz	50
Sym. short circuit current	kA	System 1: 8,1; System 2: 14,7
Short circuit duration max.	s	System 1: 2; System 2: 2
Weight of oil	t	approx. 6,5
Total weight	t	approx. 28,5

Data of instrument current transformer

Phase	2V
Prim. rated current A	1400
Sec. rated current A	2
Rated power VA	10
Accuracy class	10P10
Connected to	1S1-1S2

I a b c d e f o

Phase	2V
Connection box of current transformer	I
a	1S1
b	1S2
c	
d	
e	
f	
Earthing connection	a

Caution!
High-tension with open secondary circuit of current transformer.

Caution!
The off-load tap changer must only be operated when the transformer is de-energized.

System 1				System 2	
Volts	Connected to	Pos.	Tap changer connects	Volt	Connected to
12075	1U, 1V, 1W	1	1U5-1U6, 1V5-1V6, 1W5-1W6	6300	2U, 2V, 2W, 2N
11788		2	1U6-1U4, 1V6-1V4, 1W6-1W4		
11500		3	1U4-1U7, 1V4-1V7, 1W4-1W7		
11212		4	1U7-1U3, 1V7-1V3, 1W7-1W3		
10925		5	1U3-1U8, 1V3-1V8, 1W3-1W8		

FIGURE A.3 The specification nameplate of a 15 MVA transformer.

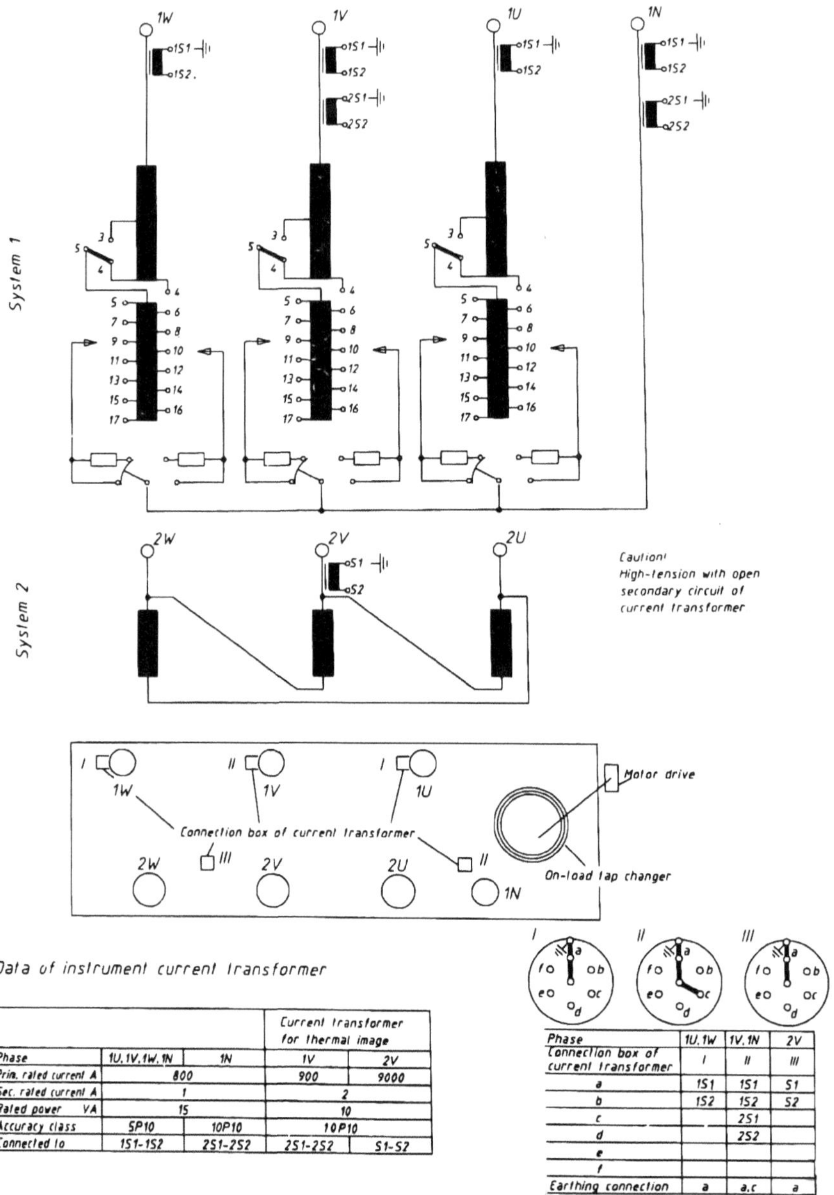

			Current transformer for thermal image	
Phase	1U, 1V, 1W, 1N	1N	1V	2V
Prim. rated current A	800		900	9000
Sec. rated current A	1		2	
Rated power VA	15		10	
Accuracy class	5P10	10P10	10P10	
Connected to	1S1-1S2	2S1-2S2	2S1-2S2	S1-S2

Phase	1U, 1W	1V, 1N	2V
Connection box of current transformer	I	II	III
a	1S1	1S1	S1
b	1S2	1S2	S2
c		2S1	
d		2S2	
e			
f			
Earthing connection	a	a, c	a

FIGURE A.4 An example of a transformer tap changer specification nameplate.

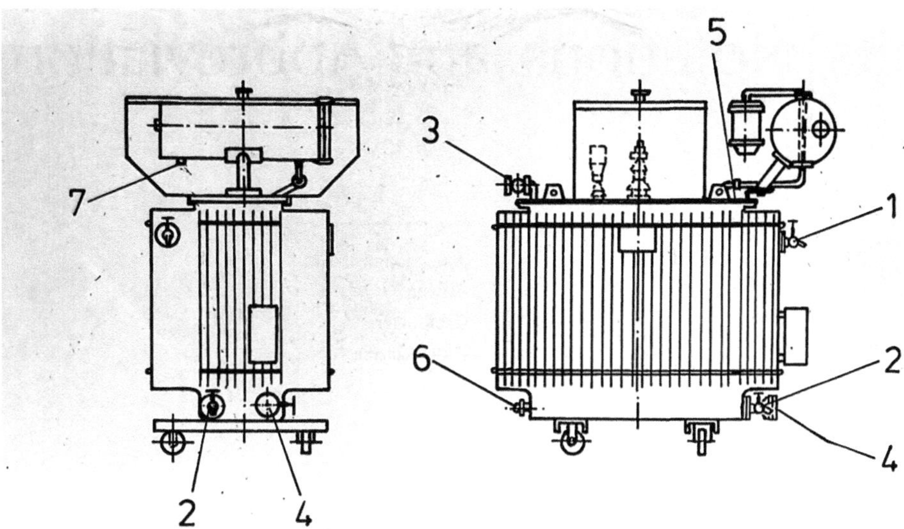

ITEM	DIMENSION	DESIGNATION	QUANTITY	NORMAL OPERATION
1	ϕ15	Oil sampling cock upper	1	closed
2	ϕ15	Oil sampling cock lower	1	closed
3	DN50	Filter valvé upper	1	closed
4	DN50	Filter valve lower	1	closed
5	R1/4"	Over pressure valve	1	closed
6	DN31	Oil drain device	1	closed
7	R1/2"	Oil drain device	1	closed

FIGURE A.5 An example of the specification nameplate of transformer oil cooling components.

Units, Notations, and Abbreviations

UNITS

AC	Alternating current
A	Ampere
cm	Centimeter
DC	Direct current
F	Farad
H	Henry
HV	High Voltage
Hz	Hertz
h	Hour
hp	Horsepower
J	Joule
kV	Kilovolt
kVA	Kilovolt Ampere
kW	Kilowatt
MW	Megawatt
MΩ	Megaohm
N	Newton
N.m	Newton meter
Ω	Ohm
pu	Perunit
rad	Radian
s	Second (time)
turn	Coil turn
V	Volt
VA	Volt-Ampere
VAr	Volt-Ampere reactive
W	Watt

NOTATION

Lowercase letters such as $v(t)$ and $i(t)$ indicate instantaneous values.
Uppercase letters in italics such as V and I indicate **rms** values.
Uppercase letters such as $\hat{V}$ and indicate **rms** phasors.
Matrices and vectors with real components such as **R** and **I** are indicated by boldface type.
Matrices and vectors with complex components such as ***Z*** and ***I*** are indicated by boldface italic type.
Superscript (t) denotes vector or matrix transpose.
Asterisk (*) denotes complex conjugate.

ABBREVIATIONS

AF	Audio-Frequency
AI	Artificial Intelligence
BDA	Big Data Analytics
CFD	Computational Fluid Dynamics
CT	Current Transformer
DAS	Distribution Automation Systems
DAU	Data Acquisition Unit
DETC	De-Energized Tap Changer
DGA	Dissolved Gas Analysis
DL	Deep Learning
EAF	Electric Arc Furnace
EMI	Electromagnetic Interference
FACTS	Flexible AC Transmission System
FEA	Finite Element Analysis
HFC	Hydrofluorocarbons
HPET	Hybrid Power Electronic Transformer
HTS	High-Temperature Superconducting Transformer
HVDC	High-Voltage Direct Current
IPFC	Interphase Power Flow Controller
LF	Ladle Furnace
ML	Machine Learning
MMC	Modular Multilevel Converter
NC	Natural Convection
NLTC	No-Load Tap Changers
OCTC	Off-Circuit Tap Changer
OFAF	Oil Forced Air Forced
OLTC	On-Load Tap Changers
ONAF	Oil Natural Air Forced
ONAN	Oil Natural Air Natural
PAR	Phase Angle Regulator
PCB	Printed Circuit Board
PD	Partial Discharge
PET	Power Electronic Transformer
PFC	Perfluorocarbons
PLC	Power Transmission Line Communication
PPE	Personal Protective Equipment
PST	Phase Shift Transformer
PT	Potential Transformer
SFRA	Sweep Frequency Response Analysis
SMPS	Switched-Mode Power Supply
SPR	Sudden Pressure Relay
SST	Solid-State Transformer
STATCOM	Static Synchronous Compensator
SVC	Static VAR Compensator
TCSC	Thyristor-Controlled Series Compensator
UPFC	Unified Power Flow Controller
VFT	Variable-Frequency Transformer
VLC	Variable Load Controller
VT	Voltage Transformer

Bibliography

When selecting a book on *Power System Transformers*, you should consider your level of expertise, the depth of information you require, and whether you prefer a more theoretical or application-oriented approach.

The following books can serve as valuable resources for students, engineers, and professionals looking to enhance their knowledge of Power System Transformers.

Below is a list of prominent books on Power System Transformers (listed without any specific order):

1. McLyman, C.W.T., *Transformer and Inductor Design Handbook (Electrical and Computer Engineering)*, 4th edition. CRC Press, Boca Raton, FL, 26 April 2011.
2. Harlow, J.H., *The Electric Power Engineering Handbook, Electric Power Transformer Engineering*, 3rd Edition. CRC Press, Boca Raton, FL, 2012.
3. Kulkarni, S.V., Khaparde, S.A., *Transformer Engineering Design Technology and Diagnostics*, 2nd edition. CRC Press, Boca Raton, FL, 6 September 2012.
4. Kothari, D.P., Nagrath, I.J., *Electric Machines*, 5th edition. Mc Graw Hill, India, 8 August 2017.
5. Zhu, F., Yang, B., *Power Transformer Design Practices*, 1st edition. CRC Press, Boca Raton, FL, 18 March 2021.
6. Vecchio, R.M.D., Poulin, B., Feghali, P.T., Shah, D.M., Ahuja, R., *Transformer Design Principles*, 3rd edition. CRC Press, Boca Raton, FL, 18 August 2017.
7. Gönen, T., Mehrizi-Sani, A., *Electrical Machines and Their Applications*, 3rd edition. CRC Press, Boca Raton, FL, 19 January 2024.

References

1. Eidiani, M., Rouzbehi, K., *Advanced Topics in Power Systems Analysis Problems, Methods, and Solutions*, 1st edition. CRC Press, Boca Raton, FL, 6 September 2024.
2. Eidiani, M., Heidari, V., *Fundamentals of Power Systems Analysis* 1*: Problems and Solutions*. Taylor & Francis Group, CRC Press, Boca Raton, FL, pp. 1–215, 2023, doi: 10.1201/9781003394433.
3. Eidiani, M., "An efficient differential equation load flow method to assess dynamic available transfer capability with wind farms," *IET Renewable Power Generation*, 2021, 15(16), pp. 3843–3855, doi: 10.1049/rpg2.12299.
4. Eidiani, M., "A new load flow method to assess the static available transfer capability," *Journal of Electrical Engineering and Technology*, 2022, 17(5), pp. 2693–2701, doi: 10.1007/s42835-022-01105-3.
5. Zeynal, H., Jiazhen, Y., Azzopardi, B., Eidiani, M., "Flexible economic load dispatch integrating electric vehicles," *2014 IEEE 8th International Power Engineering and Optimization Conference (PEOCO2014)*, Wuhan, China, pp. 520–525, 2014, doi: 10.1109/PEOCO.2014.6814484.
6. Zeynal, H., Zadeh, A.K., Nor, K.M., Eidiani, M., "Locational marginal price (LMP) assessment using hybrid active and reactive cost minimization," *International Review of Electrical Engineering*, 2010, 5(5), pp. 2413–2418.
7. Eidiani, M., Zeynal, H., Shaaban, M., "A detailed study on prevailing ATC methods for optimal solution development," *2022 IEEE International Conference on Power and Energy (PECon)*, Langkawi, Kedah, Malaysia, pp. 299–303, 2022, doi: 10.1109/PECon54459.2022.9988775.
8. Eidiani, M., "A reliable and efficient method for assessing voltage stability in transmission and distribution networks," *International Journal of Electrical Power and Energy Systems*, 2011, 33(3), pp. 453–456, doi: 10.1016/j.ijepes.2010.10.007.
9. Eidiani, M., "A new method for assessment of voltage stability in transmission and distribution networks," *International Review of Electrical Engineering*, 2010, 5(1), pp. 234–240.
10. Eidiani, M., Ashkhane, Y., Khederzadeh, M., "Reactive power compensation in order to improve static voltage stability in a network with wind generation," *2009 International Conference on Computer and Electrical Engineering, ICCEE 2009*, Dubai, United Arab Emirates, pp. 47–50, 2009, doi: 10.1109/ICCEE.2009.239.
11. Eidiani, M., Buygi, M.O., Ahmadi, S., "CTV, complex transient and voltage stability: A new method for computing dynamic ATC," *International Journal of Power and Energy Systems*, 2006, 26(3), pp. 296–304, doi: 10.2316/Journal.203.2006.3.203-3597.
12. Eidiani, M., Badokhty, M.E., Ghamat, M., Zeynal, H., "Improving transient stability using combined generator tripping and braking resistor approach," *International Review on Modelling and Simulations*, 2011, 4(4), pp. 1690–1699.
13. Eidiani, M., Baydokhty, E., Ghamat, M., Zeynal, H., Mortazavi, H., "Transient stability enhancement via hybrid technical approach," *2011 IEEE Student Conference on Research and Development*, Cyberjaya, Malaysia, pp. 375–380, 2011, doi: 10.1109/SCOReD.2011.6148768.
14. Eidiani, M., Shanechi, M.H.M., Vaahedi, E., "Fast and accurate method for computing FCTTC (first contingency total transfer capability)," *Proceedings. International Conference on Power System Technology*, Kunming, China, pp. 1213–1217, vol. 2, 2002, doi: 10.1109/ICPST.2002.1047595.
15. Zeynal, H., Hui, L.X., Jiazhen, Y., Eidiani, M., Azzopardi, B., "Improving Lagrangian relaxation unit commitment with cuckoo search algorithm," *2014 IEEE International Conference on Power and Energy (PECon)*, Kuching, Malaysia, pp. 77–82, 2014, doi: 10.1109/PECON.2014.7062417.
16. Eidiani, M., Zeynal, H., Zadeh, A.K., Mansoorzadeh, S., Nor, K.M., "Voltage stability assessment: An approach with expanded Newton Raphson-Sydel," *2011 5th International Power Engineering and Optimization Conference*, Shah Alam, Malaysia, pp. 31–35, 2011, doi: 10.1109/PEOCO.2011.5970424.
17. Eidiani, M., "Assessment of voltage stability with new NRS," *2008 IEEE 2nd International Power and Energy Conference*, Johor Bahru, Malaysia, pp. 494–496, 2008, doi: 10.1109/PECON.2008.4762525.
18. Eidiani, M., Yazdanpanah, D., "Minimum distance, a quick and simple method of determining the static ATC," *Journal of Electrical Engineering*, 2011, 11(2), pp. 16, 95–101.
19. Eidiani, M., Asadi, S.M., Faroji, S.A., Velayati, M.H., Yazdanpanah, D., "Minimum distance, a quick and simple method of determining the static ATC," *2008 IEEE 2nd International Power and Energy Conference*, Johor Bahru, Malaysia, pp. 490–493, 2008, doi: 10.1109/PECON.2008.4762524.

20. Eidiani, M., "A rapid state estimation method for calculating transmission capacity despite cyber security concerns," *IET Generation, Transmission and Distribution*, 2023, 17, pp. 1–9, doi: 10.1049/gtd2.12747.
21. Eidiani, M., "A reliable and efficient holomorphic approach to evaluate dynamic available transfer capability," *International Transactions on Electrical Energy Systems*, 2021, 31(11), e13031, pp. 1–14, doi: 10.1002/2050-7038.13031.
22. Eidiani, M., Zeynal, H., Zadeh, A.K., Nor, K.M., "Exact and efficient approach in static assessment of available transfer capability (ATC)," *2010 IEEE International Conference on Power and Energy*, Kuala Lumpur, Malaysia, pp. 189–194, 2010, doi: 10.1109/PECON.2010.5697580.
23. Eidiani, M., Zeynal, H., "A fast holomorphic method to evaluate available transmission capacity with large scale wind turbines," *9th Iranian Conference on Renewable Energy & Distributed Generation (ICREDG)*, Mashhad, Iran, pp. 1–5, 2022, doi: 10.1109/ICREDG54199.2022.9804527.
24. Eidiani, M., Zeynal, H., Zakaria, Z., "An efficient holomorphic based available transfer capability solution in presence of large scale wind farms," *2022 IEEE International Conference in Power Engineering Application (ICPEA)*, Shah Alam, Malaysia, pp. 1–5, 2022, doi: 10.1109/ICPEA53519.2022.9744711.
25. Eidiani, M., Shanechi, M.H.M., "FAD-ATC: A new method for computing dynamic ATC," *International Journal of Electrical Power and Energy Systems*, 2006, 28(2), pp. 109–118, doi: 10.1016/j.ijepes.2005.11.004.
26. Eidiani, M., "ATC evaluation by CTSA and POMP, two new methods for direct analysis of transient stability," *IEEE/PES Transmission and Distribution Conference and Exhibition*, 2002, 3, pp. 1524–1529, doi: 10.1109/TDC.2002.1176824.
27. Eidiani, M., Zeynal, H., Zakaria, Z., "Development of online dynamic ATC calculation integrating state estimation," *2022 IEEE International Conference in Power Engineering Application (ICPEA)*, Shah Alam, Malaysia, pp. 1–5, 2022, doi: 10.1109/ICPEA53519.2022.9744694.
28. Eidiani, M., Zeynal, H., Zakaria, Z., "An efficient method for available transfer capability calculation considering cyber-attacks in power systems," *2023 IEEE 3rd International Conference in Power Engineering Applications (ICPEA),* Putrajaya, Malaysia, pp. 127–130, 2023, doi: 10.1109/ICPEA56918.2023.10093168.
29. Eidiani, M., Ghavami, A., "New network design for simultaneous use of electric vehicles, photovoltaic generators, wind farms and energy storage," *2022 9th Iranian Conference on Renewable Energy & Distributed Generation (ICREDG)*, Mashhad, Iran, pp. 1–5, 2022, doi: 10.1109/ICREDG54199.2022.9804534.
30. Eidiani, M., Zeynal, H., "An effective method to determine the available transmission capacity with variable frequency transformer," *International Transactions on Electrical Energy Systems*, 2023, 2023, pp. 1–10, doi: 10.1155/2023/8404284.
31. Eidiani, M., "A new hybrid method to assess available transfer capability in AC–DC networks using the wind power plant interconnection," *IEEE Systems Journal*, 2023, 17(1), pp. 1375–1382, doi: 10.1109/JSYST.2022.3181099.
32. Eidiani, M., Zeynal, H., Zakaria, Z., "A comprehensive study on the renewable energy integration using DIgSILENT," *2023 IEEE 3rd International Conference in Power Engineering Applications (ICPEA),* Putrajaya, Malaysia, pp. 197–201, 2023, doi: 10.1109/ICPEA56918.2023.10093153.
33. Eidiani, M., Zeynal, H., Zakaria, Z., Shaaban, M., "Analysis of optimization methods applied for renewable energy integration," *2023 IEEE 3rd International Conference in Power Engineering Applications (ICPEA),* Putrajaya, Malaysia, p. 1570861924, 6–7 March 2023.
34. Eidiani, M., Applying optimization techniques to develop a renewable energy supply map. In: Fathi, M., Zio, E., Pardalos, P.M. (eds.), *Handbook of Smart Energy Systems.* Springer, Cham, 2022, doi: 10.1007/978-3-030-72322-4_61-1.
35. Eidiani, M., Integration of renewable energy sources. In: Fathi, M., Zio, E., Pardalos, P.M. (eds.), *Handbook of Smart Energy Systems.* Springer, Cham, 2022, doi: 10.1007/978-3-030-72322-4_41-1.
36. Eidiani, M., Kargar, M., "Frequency and voltage stability of the microgrid with the penetration of renewable sources," *2022 9th Iranian Conference on Renewable Energy & Distributed Generation (ICREDG)*, Mashhad, Iran, pp. 1–6, 2022, doi: 10.1109/ICREDG54199.2022.9804542.
37. Eidiani, M., Zeynal, H., Ghavami, A., Zakaria, Z., "Comparative analysis of mono-facial and bifacial photovoltaic modules for practical grid-connected solar power plant using PVsyst," *2022 IEEE International Conference on Power and Energy (PECon)*, Langkawi, Kedah, Malaysia, pp. 499–504, 2022, doi: 10.1109/PECon54459.2022.9988872.

38. Eidiani, M., Asghari Shahdehi, N., Zeynal, H., "Improving dynamic response of wind turbine driven DFIG with novel approach," *2011 IEEE Student Conference on Research and Development*, Cyberjaya, Malaysia, pp. 386–390, 2011, doi: 10.1109/SCOReD.2011.6148770.
39. Eidiani, M., Mahnani, S., "Optimally and independent planned microgrid with solar-wind and biogas hybrid renewable systems by HOMER," *Majlesi Journal of Electrical Engineering*, 2023, 17(2), pp. 153–158, doi: 10.30486/mjee.2023.1974843.1021.
40. Zeynal, H., Jiazhen, Y., Azzopardi, B., Eidiani, M., "Impact of electric vehicle's integration into the economic VAr dispatch algorithm," *2014 IEEE Innovative Smart Grid Technologies - Asia (ISGT ASIA)*, Kuala Lumpur, Malaysia, pp. 780–785, 2014, doi: 10.1109/ISGT-Asia.2014.6873892.
41. Zeynal, H., Eidiani, M., "Hydrothermal scheduling flexibility enhancement with pumped-storage units," *2014 22nd Iranian Conference on Electrical Engineering (ICEE)*, Tehran, Iran, pp. 820–825, 2014, doi: 10.1109/IranianCEE.2014.6999649.
42. Eidiani, M., Zeynal, H., "Determination of online DATC with uncertainty and state estimation," *2022 9th Iranian Conference on Renewable Energy & Distributed Generation (ICREDG)*, Mashhad, Iran, pp. 1–6, 2022, doi: 10.1109/ICREDG54199.2022.9804581.
43. Zeynal, H., Eidiani, M., Yazdanpanah, D., "Intelligent substation automation systems for robust operation of smart grids," *2014 IEEE Innovative Smart Grid Technologies - Asia (ISGT ASIA)*, Kuala Lumpur, Malaysia, pp. 786–790, 2014, doi: 10.1109/ISGT-Asia.2014.6873893.
44. Zand, Z., Ghahri, M.R., Majidi, S., Eidiani, M., Nasab, M.A., Zand, M., Smart grid and resilience. In: Fathi, M., Zio, E., Pardalos, P.M. (eds.), *Handbook of Smart Energy Systems*. Springer, Cham, 2022. doi: 10.1007/978-3-030-72322-4_178-1.
45. Momen, S., Hekmati, A., Majidi, S., Zand, Z., Zand, M., Nikoukar, J., Mostafa, E., Energy harvesting for smart energy systems. In: Fathi, M., Zio, E., Pardalos, P.M. (eds.), *Handbook of Smart Energy Systems*. Springer, Cham, 2022, doi: 10.1007/978-3-030-72322-4_12-1.
46. Baydokhty, M.E., Eidiani, M., Zeynal, H., Torkamani, H., Mortazavi, H., "Efficient generator tripping approach with minimum generation curtailment based on fuzzy system rotor angle prediction," *Przeglad Elektrotechniczny*, 2012, 88(9A), pp. 266–271.
47. Ghardashi, G., Gandomkar, M., Majidi, S., Eidiani, M., Dadfar, S., "Accuracy and speed improvement of microgrid islanding detection based on PV using frequency-reactive power feedback method," *2022 International Conference on Protection and Automation of Power Systems (IPAPS)*, Zahedan, Iran, pp. 1–8, 2022, doi: 10.1109/IPAPS55380.2022.9763190.
48. Eidiani, M., Zeynal, H., "New approach using structure-based modeling for the simulation of real power/frequency dynamics in deregulated power systems," *Turkish Journal of Electrical Engineering and Computer Sciences*, 2014, 22(5), pp. 1130–1146, doi: 10.3906/elk-1208-90.
49. Eidiani, M., "The effect of power system strength on the calculation of available transmission capacity", *Power System Strength: Evaluation methods, best practice, case studies, and applications*, 2024, pp. 137–174, doi: 10.1049/PBPO247E_ch7.
50. Eidiani, M., Modeling renewable energy resources using DIgSILENT PowerFactory software, Power Systems Operation with 100% *Renewable Energy Sources*, 2023, pp. 165–202
51. Eidiani, M. Online dynamic ATC computation with large-scale wind farms. *Electrical Engineering*, 2024. doi: 10.1007/s00202-024-02325-8
52. Eidiani, M., Zeynal, H., Ghavami, A., Zakaria, Z., "Performance of bifacial solar system for 1MW grid-connected solar power plant in Mashhad, Iran: Design," *Simulation and Economic Analysis", 2024 9th International Conference on Technology and Energy Management, ICTEM 2024*, 2024
53. Eidiani, M., Ghavami, A., Zeynal, H., Zakaria, Z., Performance of monofacial and bifacial solar system for 5MW grid-connected solar power plant with fixed tilt and tracking: design, simulation and investigating the effect of temperature", *2024 28th International Electrical Power Distribution Conference, EPDC 2024*, 2024.
54. Eidiani, M., Kargar, M., Zeynal, H., "Interactive use of D-STATCOM and storage resource to maintain microgrid stability for commercial systems", *Microgrids for Commercial Systems*, 2024, pp. 241–270.

Index

For Product Safety Concerns and Information please contact our EU representative GPSR@taylorandfrancis.com Taylor & Francis Verlag GmbH, Kaufingerstraße 24, 80331 München, Germany

Batch number: 10399852

Printed by Printforce, the Netherlands